T0135485

Physical Characterization of Coatings - Introduction to Rheology and Surface Analysis

Michael Osterhold

English Edition of the 2nd German Edition
Physikalische Charakterisierung von Lacken –
Einführung in Rheologie und Oberflächenanalyse

Bibliographic information published by the
Deutsche Nationalbibliothek

The Deutsche Nationalbibliothek lists this publication in the Deutsche Nationalbibliografie; detailed bibliographic data are available on the Internet at http://dnb.d-nb.de.

ISBN 978-3-8325-5049-3

Logos Verlag Berlin GmbH
Comeniushof, Gubener Str. 47,
10243 Berlin, Germany
Tel.:   +49 (0)30 42 85 10 90
Fax: +49 (0)30 42 85 10 92
INTERNET: http://www.logos-verlag.de

Physical Characterization of Coatings
Introduction to Rheology and Surface Analysis

Michael Osterhold

Content

Preface to the Second German Edition

Compared to the first edition, the second edition of this book has been extended by introductory chapters on rheology and thermal analysis. At the end of the book there is also a short introduction to weathering testing. The chapters on scratch resistance and surface structures have been revised and updated to emphasize the more introductory and summarizing nature of each section.

Michael Osterhold, July 2019

Preface to the First German Edition

In this book, eight chapters discuss two important areas of physical characterization of coatings. Chapters 1 – 4 deal with the characterization of the liquid coating under the keyword rheology. In addition to measuring methods for yield point determination, thixotropy and oscillation, examples of application in connection with performance properties are also described. In addition, topics such as surface charge and particle size determination are addressed. Chapters 5 – 8 under the heading surface analysis cover topics such as surface structure and methods for characterizing the scratch resistance of the cured coating film. The importance of surface tension for certain coating properties and a first insight into the analysis of defects are also presented. The individual chapters are partly based on review articles by the author in journals or on conference papers from the last 20 years. These are intentionally introductory articles for the interested readership, special knowledge in the respective fields is not required for understanding. The bibliographic references can be found in the literature sources.

Michael Osterhold, January 2018

Chapter 1 – Introduction Rheology

Flow Behaviour and Rotational Rheometers

1 Introduction

The term "viscosity" has a special connotation in the coatings industry. Many application properties such as ease of processing, sagging and levelling are influenced by "viscosity", in other words the coating's flow behaviour. This means that consistently high product quality can only be guaranteed by an exact knowledge of the coating's rheological behaviour, i. e. its flow characteristics. The unequivocal rheological characterization of the materials used requires flow curves to be produced with rotational rheometers.

2 Rotational rheometers

The various rotational rheometers on the market are based on two different measuring principles (rate-controlled or stress-controlled).

In the case of rate-controlled rotational rheometers (CR) the material is subjected to shear in a gap and the resulting shear stress is measured. This is done e. g. by using coaxial measuring systems by rotation of the outer cylinder of a Couette system (or inner cylinder of a Searle system). The shear rate for any rotational speed can be calculated from a knoweledge of the test equipment geometry. Fig. 1 shows a coaxial cylinder measuring system consisting of outer (left) and inner cylinder (right). In a so-called cone-plate measuring system (fig. 2), the shearing occurs in the gap that is formed between a plate and the cone.

Fig. 1: Coaxial cylinder measuring system, left outer cylinder, right inner cylinder

Fig. 2: Cone-plate measuring system, cone angle 1°

3 Flow behaviour

These measuring geometries for rotational rheometers are described in various DIN standards (e. g. DIN 53019, DIN Paperback 398 Rheology). The recording of a flow curve (shear stress vs. shear rate) finally allows to describe the investigated material with regard to its flow behaviour. Shear stress τ and shear rate $\dot{\gamma}$ are linked via the relationship

$$\tau = \eta \; \dot{\gamma}$$

η is the (dynamic) viscosity in this respect, and, for non-Newtonian fluids, also 'apparent' viscosity.

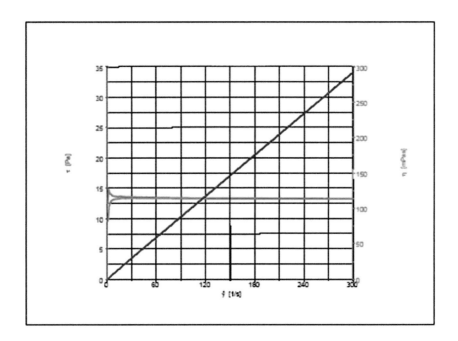

Fig. 3: Newtonian flow behaviour

Rotational rheometers must always be used for the rheological characterization of a substance, if it is not certain whether a purely Newtonian
flow behaviour is present. A Newtonian flow behaviour is characterized
by a constant viscosity independent of the shear (shear rate). Fig. 3
(acc. [3]) shows a typical flow and viscosity curve of a Newtonian min-
eral oil up to a shear rate of 300 s^{-1} (blue flow curve, green viscosity
curve).

If Newtonian flow behaviour is present, the viscosity can also be determined with the aid of flow cups (e. g. according to DIN EN ISO
2431). In contrast, purely Newtonian behaviour cannot be observed
practically in modern, in particular waterborne coating systems. Rather,
flow anomalies such as shear thinning (pseudo-plasticity), thixotropy or
even viscoelastic behaviour occur.

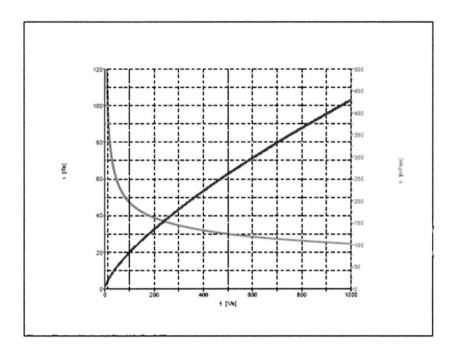

Fig. 4: Pseudo-plastic flow behaviour

If a material has a pseudo-plastic flow behaviour, the viscosity decreases with increasing shear (shear rate), in English usage it is referred to as 'shear thinning'. Fig. 4 (acc. [3]) shows flow and viscosity curves of a pseudo-plastic coating up to a shear rate of 1000 s^{-1}.

In the phenomenon of thixotropy, a temporal dependence of the viscosity is also observed (see next chapter). In contrast, the case of increasing viscosity with increasing shear (dilatant flow behaviour or rheopexy) is extremely rare for coating systems. Thixotropy and the meaning of yield points will be discussed in the next chapter.

In the case of stress-controlled rotational rheometers (CS) it is not the shear rate that is specified as in CR instruments, but the shear stress.

Here one determines deformation (rotation) and thus the shear rate. Extremely small shear rates can be realized with stress-controlled instruments, so that these instruments are especially useful for accurately determining small yield points (< 1 Pa). For a dynamic experiment (oscillation, see next chapter) typical deformations vary between < 1 % and a few 10 %; the oscillation frequency range normally varies between about 0.001 Hz and a few 10 Hz. Often used is the oscillation test with a fix frequency of 1 Hz.

An extensive general representation of rheology and rheometry can be found, for example, in [5]. It exists a detailed collection of DIN standards [2] on the usual methods of viscosity measurement.

References

01.   DIN 53019

02.   DIN-Taschenbuch 398, Rheologie, Beuth (2019)

03.   M. Osterhold, Vortragstagung der GDCh-Fachgruppe Lackchemie, Paderborn (2016)

04.   DIN EN ISO 2431

05.   T. Mezger, The Rheology Handbook, Vincentz, Hannover, Germany, 4. Edition (2014)

Chapter 2 – Rheological Measurements

Rheological Characterization of Coatings Yield Point, Thixotropy and Oscillation

1 Introduction

Many applicational and technological properties are influenced by the flow behaviour of the coating. High product quality can only be guaranteed through an exact knowledge of the rheological behaviour of the coating and the used raw materials, respectively. In view of the increasing use of waterborne systems flow anomalies as thixotropy, yield points or also viscoelastic behaviour can be observed more often. Such behaviour is not normally observed in conventional, solvent-borne coatings. If, however, so-called SCA agents (Sagging Control Agents) are added to directly control rheological properties, phenomena like thixotropy, yield points or viscoelasticity can appear as well [1-3].

Yield point and thixotropy influence important materials properties as storage stability, pumping behaviour or levelling and flowing. Against this background the Working Group "Rheology" of the Standards Committee coatings and coating materials (NAB) within the DIN (German Institute of Standardization e. V.) has followed up intensively on the measuring characterization of yield points and thixotropy in the last years. Two technical reports on these items have been prepared [3, 8].

The measuring possibilities of characterizing the rheological properties with rotational rheometers concerning yield point and thixotropy will be presented in this chapter.

2 Definition and importance of yield point and thixotropy

The yield point is defined as the lowest shear stress above which the behaviour of a material, in rheological respect, is like that of a liquid; below the yield point its behaviour is like that of an elastic or viscoelastic body.

Thixotropy describes a flow behaviour, where the rheological parameters (viscosity) decrease due to a mechanical constant load to a timely constant limiting value and after reducing the load, the initial state is completely reached depending on time. In practice, only a limited time frame is considered in which the initial state is not always reached.

With yield point and thixotropy important material properties can be characterized, e. g.

- Effectiveness of rheological additives
- Storage stability (e.g. against sedimentation, separation, flocculation)
- Behaviour when starting to pump
- Wet film thickness
- Levelling and flowing behaviour
  (e. g. without brushmarks and sag formation)
- Orientation of effect pigments

3 Methods for determining the yield point

The individual methods for determining the yield point are summerized and critically discussed in the DIN technical report 143 [4]. The presented results for evaluating yield points in this report base on interlaboratory tests, which were carried out by the participants of the Working Group "Rheology" of the Standards Committees "Pigments and Extenders" and "Coatings and Coatings Materials" at DIN.

In first preliminary tests different waterborne basecoats with low and dispersions with distinctly higher yield points have been examined as well. It was found that some methods showed unexpectedly good qualitative relationships. On the other hand, some participants reported problems with the preparation of the samples. In addition, in the course of the preliminary tests certain methods of measurement have been found to be unsuitable for the samples examined and were therefore no longer considered. In this connection, the method of maximum viscosity and the method using a vane measuring system have to be mentioned. Also the method for determining the yield point using a linear stress ramp was not helpful as there are not enough measuring points in the lower measuring range. Also evaluation procedures based on traditional regression methods (e. g. according to Bingham or Herschel-Bulkley) were not considered in further tests. The results depend too strongly on the theoretical model used and the measuring specifications (ramps) (according to [3, 4]).

3.1 Tangent method in a representation of a lg γ / lg τ diagram

Based on the experiences made in preleminary tests, the participants agreed on the method "stress ramp" to be used in a continued interlaboratory test. The results are presented in the technical report 143. Therefore, below and above the assumed yield point one decade for the evaluation should be available. The (logarithmic) shear stress ramp should begin at least one decade below the assumed yield point and should reach at least one decade beyond the yield point value.

Exact test conditions have been agreed by the Working Group and definitely been specified for all participants of the comparative testing programme (see [4]). Five different samples have been examined in total: two waterborne basecoats with low yield points of a magnitude of 1 Pa or smaller, two dispersions and one sample with well known yield point provided by the Physikalisch-Technischen Bundesanstalt (PTB, The National Metrology Institute of Germany).

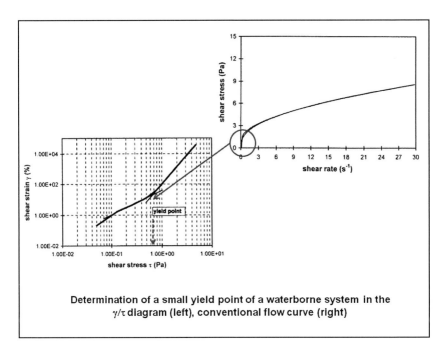

Determination of a small yield point of a waterborne system in the
γ/τ diagram (left), conventional flow curve (right)

Fig. 1: Determination of the yield point

If a yield point is existing, a straight line can be observed in the range of low shear stress, shear stress τ and shear deformation/strain γ are then proportional at low values. The investigated material shows consequently a reversible linear-elastic deformation behaviour (Hooke's elasticity law). At higher stresses the structure at rest breaks down, the deformation then becomes disproportionately high, and the material shows irreversible viscoelastic or viscous flow behaviour [5-7]. The yield point has been exceeded if the measuring points are no longer on a straight line ("tangent"). If it is possible to set up a second tangent through the measuring points at high deformations – also in the flow range – , then the crossover point of the tangents is evaluated as yield point (fig. 1) [3]. This evaluation method is described in [4] in more detail.

In summary, good results have been obtained which was due to the specifed time schedule of the tests and the previously specified measuring conditions.

For a clear characterization of yield points for various products, detailed test specifications for different substance classes have to be developed.

4 Measuring characterization of thixotropy

After finishing the investigations and the preparation of the technical report on yield point determination, 2005 first tests concerning "Thixotropy" were initiated by the DIN Working Group "Rheology". At the beginning, the target was to prove the fundamental suitability, in a second comparative testing programme (interlaboratory tests) in 2009/2010 special emphasis was given to the determination of precision data for different measuring procedures. For that purpose a waterborne basecoat and a clearcoat were investigated with four different measuring methods. A Newtonian liquid of the PTB was used as a reference. The final technical report on thixotropy was published in September 2012 [8].

A widely spread method for determining the thixotropy bases on the registration of flow curves with defined parameters for the measuring procedure. To fix are the time in which a certain maximal shear rate has to be reached (ramp time for up-curve, number of measuring points etc.), holding time at max. shear rate and time for the down-curve. Furthermore, it is to decide, wether a linear or logarithmic ramp has to be used (continuously or step-like). Besides a precise temperature control, a waiting time or pre-shearing just before the actual measurement could be important. This procedure by means of flow curves is called hysteresis method or thixotropic loop. Here, the area between up- and down-curve is evaluated as a measure for the grade of thixotropy.

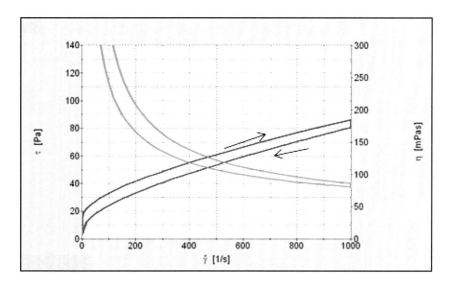

Fig. 2: Flow curve (red) and viscosity curve (green) of a blue metallic waterborne basecoat

In fig. 2 the flow curve (red) of a blue waterborne basecoat is shown as a typical example of a thixotropic coating system. The added arrows indicate each the direction of increasing (up-curve →) and decreasing (down-curve ←) shear rate, respectively. A marked hysteresis as an indicator of thixotropy can be clearly detected. In addition, two marked yield points can be observed in the low shear rate range. Also shown is the respective (apparent) viscosity run (green curves). The viscosity η decreases for higher shear rates $\dot{\gamma}$ to a value of approx. 80 mPas at

1000 s⁻¹ , and after reducing the load, i. e. at low shear rates, to increase again.

Fig. 3 shows the comparison of measurements of a waterborne basecoat with linear and logarithmic ramp [8]. Marked differences can be observed for the hysteresis areas.

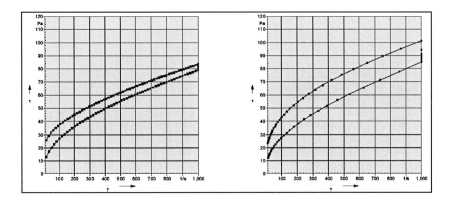

Fig. 3: Flow curves of a waterborne basecoat (left linear, right logarithmic ramp)

Due to the increasing development of controlled-stress rotational rheometers in the last years, combinations of shear and/or oscillation procedures are applied. Such experiments are usually divided into three segments (see also fig. 4 [8]). In this figure, the preset profile and the measuring results are schematically shown for the rotational case. In a first step the sample under investigation is subjected to a low shear rate (rotation), oscillation or shear stress, followed by a severe loading under rotation with high shear rate, and finally the recovery phase at a low load under rotation (shear rate), oscillation or shear stress (recovery/structure re-build).

In the second technical report of the DIN working panel several items concerning the determination of thixotropy are discussed in detail [8].

Rheological methods and the results of comparative testings of up to nine different laboratories are presented.

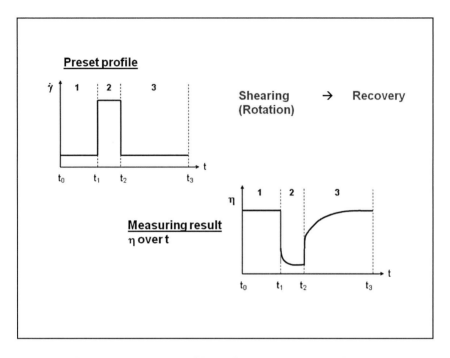

Fig. 4: Schematical preset profile and measuring result for rotation: time-dependent viscosity function of a thixotropic substance, (1) at low, (2) at high shear load with reduction of structure, and (3) again at low shear load with recovery of structure

The results show, that the method for the determination of thixotropy using flow curves (hysteresis area) is only partly suitable. Methods combining low and severe shear loads followed by the structure recovery phase at a low load allow a better evidence concerning thixotropy.

Considering values in the phase of structure recovery (3rd segment) related to values in the 2nd segment (high shearing), suitable characteristic values like TI (thixotropy index) or SRI (structure recovery index) can be determined. Precise evaluation methods and proposals for carrying out the measurements as well as presets for the measuring parameters are descript in detail in [8].

5 Oscillation measurements

Rather than using a conventional stationary shear (i. e. continuous rotation of the outer cylinder of a coaxial system), the samples were subjected to dynamic measurements in which the outer cylinder oscillated at a frequency ν ($\omega=2\pi\nu$). For a cone and plate system the oscillation of the cone or plate causes periodic deformation of the sample. The resulting force is measured using a torque transducer.

For a viscoelastic material the (shear) force and (shear) strain are not in phase, but instead the deformation lags the force by a time quantified as the phase shift. Knowledge of the force or shear stress σ, the strain γ and the phase shift δ allows one to completely describe the viscoelastic properties of the material using two parameters, $G'(\omega)$ and $G''(\omega)$ [9].

The storage modulus

$$G' = \left(\frac{\sigma_0}{\gamma_0}\right) \cos \delta$$

is a measure of the material's elasticity (where σ_0 and γ_0 are the shear stress and shear strain amplitudes, respectively).

The loss modulus

$$G'' = \left(\frac{\sigma_0}{\gamma_0}\right) \sin \delta$$

is a measure of the material's dissipation, i. e. the amount of deformation energy converted to heat. The limiting cases of $\delta=0°$ and $\delta=90°$ correspond to purely elastic and purely viscous behaviour, respectively. These moduli are components of the complex shear modulus, defined as $G^*=G'+iG''$, $|G^*|=(G'^2+G''^2)^{1/2}$.

The ratio

$$\frac{G''}{G'} = \tan \delta$$

is a measure of the ratio of energy liberated as heat to energy stored by the material. From the complex shear modulus G^* one can calculate the complex viscosity

$$|\eta^*| = \frac{|G^*|}{\omega}$$

where

$$\eta' = \frac{G''}{\omega}, \qquad \eta'' = \frac{G'}{\omega}$$

The empirical Cox–Merz rule relates the complex viscosity $|\eta^*|$ at the frequency ω to the dynamic viscosity η at the shear rate $\dot{\gamma}$

$$|\eta^*(\omega)| = \eta(\dot{\gamma}) \quad \text{for } \omega = \dot{\gamma}$$

The Cox–Merz rule has been experimentally verified for various polymer systems [10]. Since viscoelastic properties depend on frequency, the viscoelastic behaviour of a sample is observed at different oscillation frequencies. These determinations are carried out at small amplitudes, thereby ensuring that the sample is not destroyed by excessive deformation amplitudes. To determine the maximum permissible deformation amplitude (linear-viscoelastic region) one uses the so-called strain sweep test, in which the deformation amplitude is gradually increased at a fixed oscillation frequency. The amplitude up to which G^*, G' and G'' are constant, is the maximum permissible amplitude for the actual oscillation measurement.

6 Summary

The yield point and thixotropy of a system can be used to characterize important material properties, such as the effectiveness of rheological additives, storage stability, or the levelling or sagging behaviour. In addition, when these parameters are linked to application properties, such as special-effect development in waterborne coatings, interesting connections can be revealed (see next chapter).

Simple, speed-controlled rotational rheometers are often used in paint development and especially in quality assurance. They are capable of recording flow curves of the materials used, with the result that a preliminary overview of flow behaviour, i. e., Newtonian, shear thinning, thixotropic, etc., can be obtained. In the case of shear stress-controlled rotational rheometers, shear stress is specified instead of shear rate. With the aid of these instruments, it is possible to generate extremely low shear and thus to determine low yield points. Many paint makers would already have these instruments in their analytical or physical departments.

The results of the round-robin experiments have shown that recording flow curves for the purpose of determining thixotropy is less than ideal for making quantitative statements (based on the hysteresis area). However, flow curves are quiet useful for gaining a first impression of the general flow behaviour of a sample. In the context of "thixotropy", however, methods based on a combination of low and high shear load, with a subsequent phase of structural regeneration, are more promising. Product-specific values need to be developed for this.

References

01.   M. Osterhold, Prog. Org. Coat., 40 (2000) 131

02.   M. Osterhold, Farbe Lack, 116 (2010), No. 9, 33

03.   M. Osterhold, Proc. DFO „Qualitätstage", Köln, Germany (2011) 99

04.   H. Bauer, E. Fischle, L. Gehm, W. Marquardt, T. Mezger and M. Osterhold, DIN-Fachbericht 143 – Moderne rheologische

Prüfverfahren – Teil 1: Bestimmung der Fließgrenze, Grundlagen und Ringversuch, Beuth-Verlag, Berlin (2005)*; and summary of the report

05.　L. Gehm, Rheologie, Vincentz, Hannover (1998)

06.　G. Schramm, Einführung in Rheologie und Rheometrie, Haake, Karlsruhe (1995)

07.　T. Mezger, Das Rheologie-Handbuch, Vincentz, Hannover (2000)

08.　E. Fischle, E. Frigge, L. Gehm, H. Klee-Wohlenberg, C. Küchenmeister, T. Mezger, M. Osterhold, T. Remmler, U. Weckenmann, H. Wolf and R. Worlitsch, DIN SPEC 91143-2 – Moderne rheologische Prüfverfahren – Teil 2: Thixotropie, Bestimmung der zeitabhängigen Strukturänderung – Grundlagen und Ringversuch, Beuth-Verlag, Berlin (2012)*

09.　J. D. Ferry, Viscoelastic Properties of Polymers, Wiley, New York (1980)

10.　T. S. R. Al-Hadithi, H.A. Barnes and K. Walters, Colloid Polym. Sci., 270 (1992) 40

*Text in German and English

Chapter 3 – Application Examples

Rheological Methods for Coating Systems

1 Introduction

Special effect coatings are being used to an increasing extent in the automotive industry where not only pure metallic pigments such as aluminium pigments but also pearlescent pigments (mica) are increasingly being employed. Coatings with metallic or pearlescent pigments exhibit a characteristic lightness flop and/or colour flop, i. e. the impression of lightness and reflected colour alter with the angle of incident light and the angle of observation. The term flop in this article means the development of the so-called light–dark flop effect. A second part presents measurements of powder and electrodeposition coatings and of clearcoats with the oscillation technique to determine the viscosity–temperature behaviour.

2 Effect development in waterborne basecoats

This part presents analyses on flop in aqueous pearlescent/coloured pigment systems. The coloured pigments (black, green and red) used in four to five different mixtures with the pearlescent pigments were a carbon black, a Cu-phthalocyanine and a quinacridone pigment. All the basecoats (sprayed and unsprayed) were analysed for a wide range of rheological properties. Moreover, the lightness values L* of painted stoved panels (waterborne basecoats with clearcoat) were measured as a function of the gloss angle difference (20 to 70°). The results were used for a quantitative assessment of the effect (lightness flop), namely the flop-index, which depends essentially on the L* values at 20° and 70°.

2.1 Method of measurement

The flop-index is defined as follows:

$$\text{Flop-index} = \frac{2.69(L_{20}^* - L_{70}^*)^{1.11}}{(L_{45}^*)^{0.86}}$$

This definition was provided in accordance with a literature citation on metallic flop development [1]. A high flop-index means a good effect.

The measurements for G' and δ (linear-viscoelastic region) were carried out with a Bohlin VOR-Rheometer (coaxial cylinders). Examples of how oscillation measurements are applied to binders and paints can be found e. g. in [2].

26

Measurement conditions:
Oscillation frequency: 1 Hz
Temperature: 23 °C
Strain: $(2–20)\times10^{-3}$

The yield point is taken to mean the limiting shear stress below which a substance behaves like a solid. Above the yield point the substance starts to flow. The yield point can be determined with CS-rheometers (controlled-stress), in this case an "RS 100" from Haake/Germany, by plotting the resulting deformation (shear strain) as a function of the increasing shear stresses (see also chapter 2).

In parallel with these tests, flow curves were recorded with a rotational viscometer (RV 100/CV 100 from Haake) up to a shear rate of 30 s^{-1} at 23°C in order to characterize the general rheological behaviour. The characteristic determined from these curves was the viscosity at 15 s^{-1}.

In order to determine the rheological parameters, the liquid paints were measured in the sprayed and unsprayed state. The procedure was as follows: firstly, aqueous automotive repair basecoats with the desired pigmentation were prepared. The basecoats were adjusted to spray viscosity according to each specific formulation using a flow cup. The basecoats were either measured directly in the form in which they were applied (unsprayed state) or first sprayed with a spray gun, collected and then measured (sprayed state).

2.2 Results of measurements

Good correlations of the rheological parameters and the flop effect were found for the sprayed materials. For example, in the case of the phase shift δ, a correlation of varying intensity is observed for the three colours which can be combined to give a clear correlation with the flop-index – irrespective of colour (fig. 1, above).

The unsprayed materials, on the other hand, do not provide any correlations.

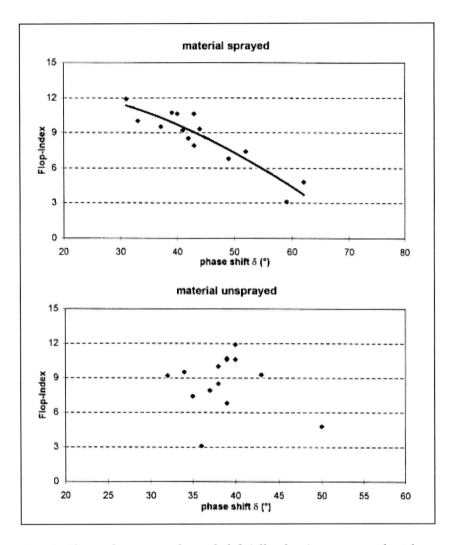

Fig. 1: Flop-index versus phase shift δ (all values), upper graph with trend line

Fig. 2 and fig. 3 show the corresponding values for storage modulus G' and yield point of the sprayed materials as a function of flop-index. Here too, good correlations are observed.

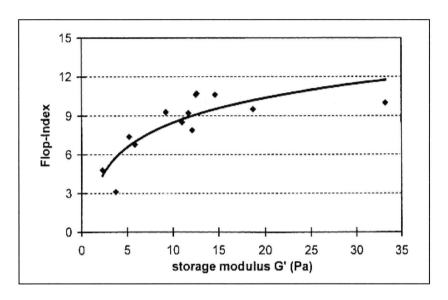

Fig. 2: Flop-index versus storage modulus G' (all values), sprayed material, with trend line

The following relationships were found (for sprayed material) [3, 4] :

• The flop-index (as a measure of effect) falls as the phase shift δ rises.

• The flop-index rises as the elastic component G' rises.

• The flop-index rises as the yield point increases.

• The correlation between the viscosity and the flop values is not so strongly pronounced.

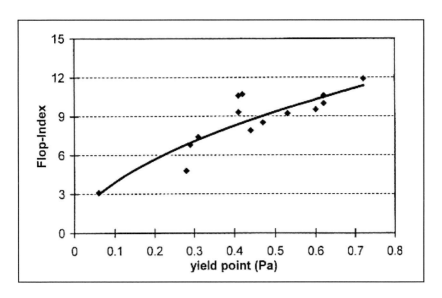

Fig. 3: Flop-index versus yield point (all values), sprayed material, with trend line

3 Measurements of viscosity-temperature behaviour

Rotational viscometers with measuring systems such as cone and plate or parallel plates can be used to characterize the rheological properties of powder and electrodeposition coatings and of clearcoats.

Using constant frequency oscillatory measurements to characterize viscosity–temperature behaviour has an advantage over continuous shear; the use of small amplitudes allows one to test materials with higher viscosities and higher solids loadings. The measurements are also less likely to be influenced by experimental effects, such as no sucking out of gap-material during shearing. The complex viscosities determined in dynamic experiments are not identical to the absolute viscosity values; however, the two can be related via the Cox–Merz rule. It is therefore referred to the measurements as "viscosities" in the remainder of this part of the chapter, rather than using the term "complex viscosity".

3.1 Powder coatings

In this section the viscosity–temperature behaviour of a powder coating for industrial applications will be presented. Measurements should be conducted at the lowest possible deformation rate to ensure that the measured rheological behaviour is similar to that which is relevant during the cure process. During the film building process only small shear rates are present, due either to running/dripping of the paint or to relative flow/melting among the individual powder coating particles [5]. The dynamic measurements presented here were obtained using a Bohlin VOR-Rheometer (Bohlin, Sweden) with a high temperature cell and cone and plate fixtures. The complex viscosity was measured at a constant frequency of 1 Hz and an amplitude of 0.2 at approximately 90-170 °C . The heating rates for the measurements were 1, 2, 4, 7, 10 and 12 K/min.

3.1.2 Viscosity level and heating rate

As evident from fig. 4, the rheological behaviour of the powder coating depends strongly on the heating rate [6]. Only the general trend is similar for the different heating rates, i. e. initial decrease in viscosity with increasing temperature until a minimum is reached, followed by increasing viscosity during cross-linking. Both the temperature at which the viscosity minimises (T_{min} at η_{min}^{*}) as well as the relative values of the viscosity differ significantly for the six heating rates.

The results can be qualitatively summarized as follows: High heating rates lead to low viscosities. Higher heating rates shift the viscosity minimum to higher temperatures. These observations have an important practical significance because to develop optimal applicability of the powder coating, it is desirable to have a low viscosity during the cross-linking phase [7, 8 und 9].

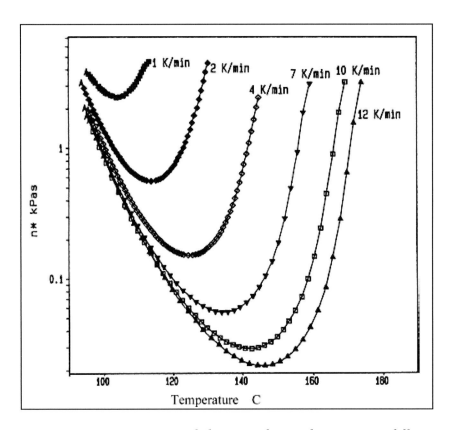

Fig. 4: Viscosity–temperature behaviour of a powder coating at different heating rates

3.1.2 Viscosity and surface structure

In order to demonstrate the relationship between the rheological behaviour of powder coatings and the surface structures of the final powder coating film, the powder coating was applied to panels and heated to 180 °C at the same heating rates as were used for the rheological measurements. They were then cured at this temperature for 5 min and measured. The coat thickness equalled approximately 90 µm. The wave-scan method (Byk-Gardner, Germany) commonly used in the automotive industry was used to survey the surface structure. The values of so-called "long-waviness" (structures approx. 10 to 1 mm) and "short-waviness" (structures < 1 mm) goes from 0 to 100, with lower values indicating a smoother structure.

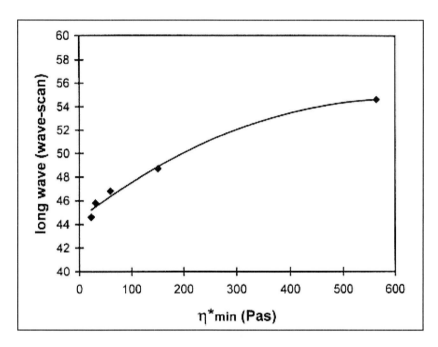

Fig. 5: Parameter 'long-wave' of the surface as a function of η_{min}^* for a powder coating

The long-wave values are shown in fig. 5 as a function of η_{min}^*. The long-waviness clearly increases in line with viscosity. The decrease in long-waviness as the heating rate increases can not only be measured with the wave-scanning equipment, but also seen with the naked eye. The sample panels cured at different heating rates were placed directly side by side and compared with one another at a flat angle looking against the daylight. A smoother surface and consequently a better flow can clearly be seen as the heating rate increases.

3.2 Electrodeposition coatings

In the following section rheological measurements are described conducted on various pigmented cathodic electrodeposition coatings; the objective was to characterize the viscosity–temperature behaviour of these coatings during the baking process concerning, which can be related to their edge covering characteristics [10].

For sample preparation a cathodic-precipitable amino epoxide resin and a cross-linking resin based on oxime-blocked hexamethylene diisocyanate were used. The pigment consisted of a carbon black–kaolin mixture with a ratio of 1:9. The coating was thinned with water to achieve a solids concentration of 15 % and then deposited onto a steel plate that was previously treated with a polysiloxane solution. The films were detached and dried for 1 h at 60 °C, after which their properties as a function of temperature were measured using an oscillatory rheometer (Bohlin VOR, cone/plate, strain 0.2, 1 Hz, 5 K/min).

The amount of pigment in the electrodeposition coating was varied based on the following binder/pigment ratios: 1:0.2, 1:0.4, and 1:0.8. The viscosity–temperature curves corresponding to these three ratios are shown in fig. 6. For each of these samples the viscosity decreases substantially with increasing temperature until reaching a minimum at 130 °C, after which it sharply increases at about 145 °C. This sharp increase results from chemical cross-linking. The influence of pigment level can be seen primarily in the increased magnitude of the viscosity with increasing pigment content. For example, the minimum viscosity values range from approximately 0.3 Pa (1:0.2) to 1 Pa (1:0.4) up to 3 Pa (1:0.8). The shape of the curves is similar for each of the samples. The minimum viscosity is directly related to the tendency of the coating to volatilise at the edges.

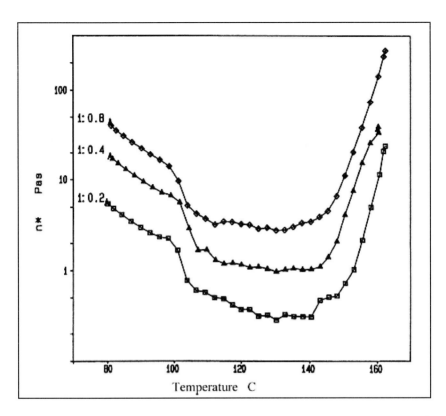

Fig. 6: Viscosity–temperature behaviour for different pigment levels of an electrodeposition coating

3.3 Clearcoats

A conventional 1 K clearcoat was studied to determine the effect of SCA (sagging control agent) on the viscosity–temperature behaviour (fig. 7, Bohlin VOR, cone/plate, strain 0.2, 1 Hz, 10 K/min). Here, for the unmodified clearcoat a nearly independent viscosity behaviour on temperature up to 120 °C can be observed, followed by a sharp increase. In contrast to that, the SCA-modified clearcoat shows a higher constant viscosity level up to 110 °C until reaching a minimum at approximately 120 °C; the comparison of the viscosity–temperature behaviour of the unmodified and the SCA-modified clearcoats can be used to optimise the levelling performances of clearcoats. For these investigations sprayed material was used (i. e. captured after spraying).

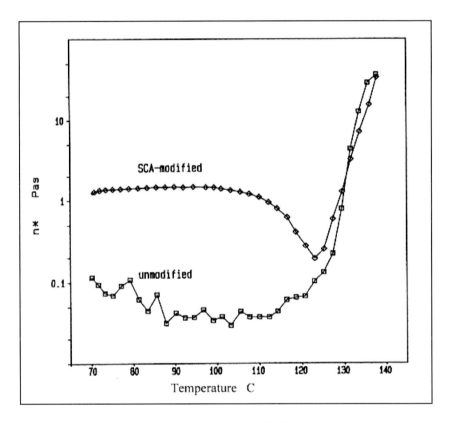

Fig. 7: Viscosity–temperature behaviour of a 1 K clearcoat

4 Summary

A first part of the chapter presented investigations on flop in aqueous automotive refinishing basecoats. All the basecoats (sprayed and un-sprayed) were analysed for a wide range of rheological properties. Moreover, the lightness values L* of painted stoved panels (waterborne basecoats with clearcoat) were measured as a function of the gloss angle difference. The results were used for a quantitative assessment of the effect (lightness flop). Good correlations of the rheological parameters and the flop effect were found for the sprayed materials. In a second part measurements concerning the viscosity–temperature behaviour of powders, electrodeposition coatings and clearcoats were demonstrated and correlated with application properties.

References

01.   A. B. J. Rodrigues, JOCCA, 4 (1992) 150

02.   M. Osterhold, Prog. Org. Coat., 40 (2000) 131

03.   M. Osterhold, G. Tönnissen, K. Wandelmaier and Th. Brock, Farbe Lack, 103 (1997) 32

04.   M. Osterhold, G. Tönnissen, K. Wandelmaier and Th. Brock, Eur. Coat. J., (1998), No. 07-08, 538

05.   M. Breucker, A. Brands, H.-D. Diesel, W. Lenhard and A.C. Bentzen, Farbe Lack, 96 (1990) 103

06.   M. Osterhold and F. Niggemann, Prog. Org. Coat., 33 (1998) 55

07.   M. Osterhold and Y. Jäger, Farbe Lack ,99 (1993) 852

08.   M. Osterhold and Y. Jäger, Rheology, 93 (1993) 250

09.   R. H. Coates, Surf. Coat. Aust., 6 (1990) 14

10.   W. Collong, M. Osterhold and Y. Voskuhl,
       Appl. Rheol. , 2 (1996) 27

Chapter 4 – Rheology and Surface Charge

Characterization of Disperse Systems (Part 1)

1 Introduction

With regard to the increasing use of modern water-based paint systems for industrial coatings, their physical characterization is becoming more and more important to guarantee a constant product quality and to develop new products. Especially parameters like flow behaviour, surface charges of the dispersed particles (e. g. pigments) or the particle size distribution have great influence on storage stability and application properties of the water-based disperse systems. For example, the stability of a pigment dispersion against flocculation is governed mainly by the interactions between the particles (surface charge), which can be greatly influenced by changes in system parameters such as pH [1]. In this chapter the relevant measuring techniques for the characterization of the physical properties (rheology, particle charge and size distribution) of the disperse systems will be presented [2]. The examples given in the chapters "Rheology" and "Surface Charge" are based on an extensive investigation concerning the influence of different amines (neutralization agents) on the stability of pigmented water-based paint systems for industrial coatings [3]. In the following paper [4] the correlation of the results with technical paint properties will be discussed.

2 Rheology

Compared to conventional solvent-based paint systems, showing often Newtonian flow behaviour, different flow behaviours like pseudoplasticity, thixotropy or also plastic behaviour can be detected for water-based paint systems [5]. In some cases viscoelastic effects can arise in non-pigmented binder dispersions, which is based on the colloidal state of the binder dispersion [6-9]. For a clear determination of the rheological behaviour flow curves are currently measured with rotational viscometers. In this case, coaxial cylinder measuring systems were mostly used, because they show some advantages compared to cone/plate-systems (e. g. no sucking out of gap material during shearing). An overview about various rheological methods is given in [10-12].

2.1 Paints

Four different paints with various pigments were investigated: carbon black, Fe-oxide, Cu-phthalocyanine (CuPc), titanium dioxide. In order to study the influence of the pH-value, the paints were adjusted with dimethylethanolamine (DMEA) to pH-values of 6.8, 7.8 and 8.8. Moreover, for the paints with the pigments carbon black and Cu-phthalocyanine an amine variation was made with ammonia (NH_3) and N-methylmorpholin (NMM) at constant pH-value 7.8. All paints contain a polyacrylic thickener.

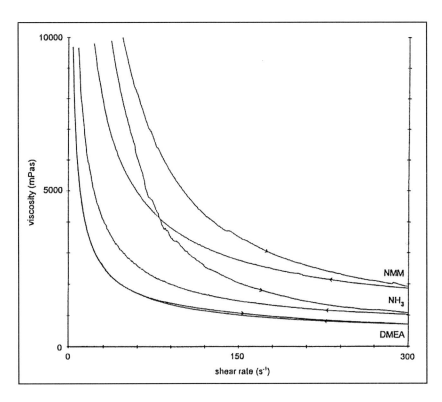

Fig. 1: Viscosity curves for carbon black paint (amine variation)

The present measurements were performed by using a rotational visco-meter RV20/CV20 with the coaxial cylinder sensor system ZA 15 (Haake Messtechnik). The measurements were run up to a maximum shear rate of 300 s^{-1} at a constant temperature of 23 °C. The ramp time was 5 min for up- and down-curve each, the holding time at maximum shear rate 1 min.

Fig. 1 exemplary shows the influence of the amine variation on the viscosity behaviour for the carbon black paint.

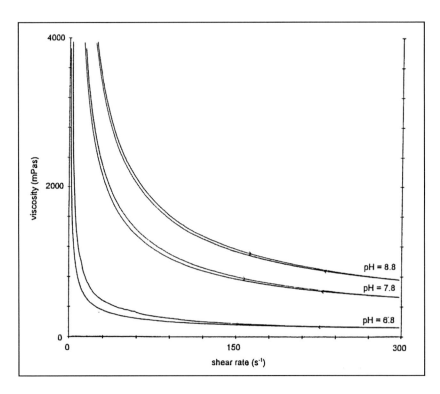

Fig. 2: Viscosity curves for CuPc paint (pH variation)

While NMM and NH_3 yield a thixotropic behaviour with high viscosity values, pseudo-plastic or rather plastic behaviour can be observed for the paint with DMEA with comparatively lower viscosity at higher shear rates. The high yield points of the down-curves are not shown in this figure. They decrease from 83 Pa (NMM) to 48 Pa (NH_3) to 22 Pa for DMEA. The yield points were estimated by fitting a rheological model (Herschel-Bulkley) to the measured flow curves. In all, the paint with DMEA shows better application properties, because pseudo-plastic materials can in general be processed more easily compared to thixotropic materials.

42

The pH variation yields e. g. for the CuPc paint an increasing viscosity with nearly pseudoplastic or rather plastic flow behaviour for increasing pH-value (Fig. 2). This confirms, that the effect of the thickener is fully developed in the alkaline pH range. The yield point increases from 5 to 47 Pa too.

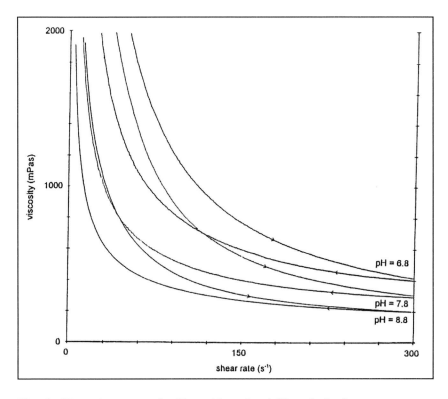

Fig. 3: Viscosity curves for Fe-oxide paint (pH variation)

In contrast to that, an opposite pH dependence with thixotropic flow behaviour can be observed for the paint with the Fe-oxide pigment (see fig. 3).

The results of the investigations can be summarized at follows: In case of amine variation the carbon black paint is distinctly more influenced in comparison to the CuPc paint. Carbon black, titanium dioxide and CuPc paints show a viscosity increase for increasing pH-value. The Fe-oxide paint responds in an inverse way.

3 Surface charge

The stability of disperse systems is governed by the interaction between the particles. If a dispersion is not sufficiently stabilized then flocculation occurs. Colloidal systems can be stabilized sterically, electrostatically or both. In water-based systems the stabilization based on electrical charges on the surfaces of the dispersed particles is dominating [13, 14]. In order to enhance the stability properties special additives are used e. g. in pigment dispersions, which should be adsorbed on the pigment surfaces and thereby stabilizing the dispersion due to their electrical charges.

The electrostatic stabilization is based on Coulomb-repulsion between the single charged particles having surface charges of the same kind (positive or negative). In an electrolyte or water the charged particles are surrounded by a fixed layer of counter-ions followed by a diffuse layer. The effective surface charge, that is real charge including the charge of the counter-ions in the fixed layer, is responsible for the electrostatical stabilization. This effective surface charge can be greatly influenced by changes in system parameters such as e. g. pH.

The stability of such a system can be described theoretically by means of the DLVO-theory [15, 16]. In literature the so-called zeta potential, that is the potential at the shear plane (transition fixed/diffuse counter-ion layer), is considered as a criterion of good dispersion stability [17]. The zeta potential is determined with the mass-transport- or micro-electrophoresis [18, 19] or also with recent techniques like the Electro-kinetic Sonic Amplitude method ESA [20, 21].

In comparison to that, in this study the effective surface charge of pigments and paints is determined directly by means of polyelectrolyte titration with a Particle Charge Detector (PCD), which works on the principle of streaming current detection. This measuring method was applied successfully for the colloid-chemical/physical characterization of latices, pigments and additives [19, 22, 23].

3.1 Polyelectrolyte titration

The measurement of the effective surface charge is performed by using a Particle Charge Detector (PCD 02, Mütek). For that purpose, the sample under investigation is filled as an aqueous slurry in a cylindric vessel made of PTFE, with a displacement piston moving periodically inside (Fig. 4). The oscillating movement of the piston causes a fluid flow, which shears off the mobile charge cloud in the surrounding of the particles, but not the fixed adsorbed layer. For this reason the charge centres of the particles, which are adsorbed on the vessel wall, and of the mobile charge cloud are shifted and a potential difference is build up along the streaming region (fig. 5). The potential difference is detected by measuring electrodes [24]. This mechanism for inducing the streaming potential is valid in the case of polymer particles. In comparison, solid particles of higher density can be hardly adsorbed on the vessel wall. Here, the induced streaming potential is based on density differences between solid particles and surrounding medium (inertia).

The actual determination of charges is carried out by means of the so-called polyelectrolyte titration. Polyelectrolytes are macromolecules with ionic groups (cationic or anionic) along their chains. The quantitative determination of the surface charge of the particles is made by adding polyelctrolytes of opposite charge until the particle surface charge is completely compensated and so the detected potential at the PCD is turned to zero. The charge concentration of the polyelectrolytes is known and by their consumption to the neutral point, the charge content of the sample can be determined.

As standard reagents for titration 0.001-molar solutions of two polyelectrolytes (Hoechst AG) were used. As anionic polyelectrolyte polyethensulfonicacid sodium (PES-Na) was used, as cationic polyelectrolyte polydiallyldimethyl-ammoniumchloride (Poly-DADMAC).

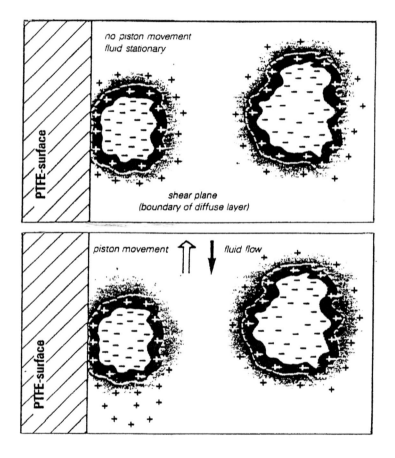

Fig. 4: Schematical representation for streaming current potential build-up (top: stationary fluid, down: moving fluid) [24]

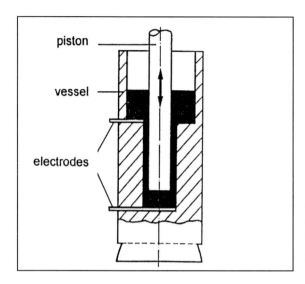

Fig. 5: Schematical set-up Particle Charge Detector [24]

Normally, before the actual measurement, pigments e. g. are dispersed at 10 % in deionised water, followed by an adjustment to three different pH-values using HCl or NaOH (0.1-molar) and then measured with the PCD.

Fig. 6 shows the measurement results of four various pigments at different pH-values. On the ordinate the respective consumption of polyelectrolyte in µmol – related to 1 g solid – is drawn (surface charge equivalent). The charge sign is opposite to that of the corresponding polyelectrolyte.

Between the single pigments marked differences can be observed. While carbon black and also Fe-oxide show especially in the alkaline range a strong negative charge, which decreases with decreasing pH-value, Cu-phthalocyanine yields only small charges. In the case of titanium dioxide a change of the charge sign can be observed at a pH-value of about 6.

Additives and thickeners show in contrast to pigments distinctly higher charges, which can be very marked especially in the alkaline range [23, 3].

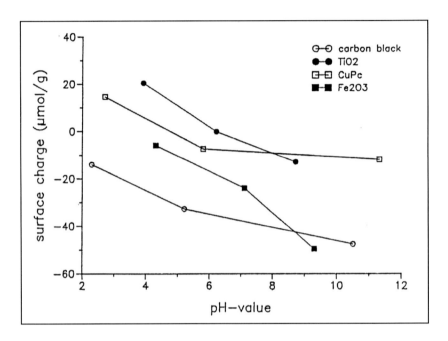

Fig. 6: Measuring results of surface charge for pigments

Surface charge measurements were performed on paints (based on the above mentioned pigments) with variation of amine type and pH. The following measurements which are presented show typical exemplary shapes. The higher charges, which can generally be observed, are based on the content of ionic binders and above all polyacrylic thickeners.

In the case of carbon black paints the amine variation leads to a small increase of negative surface charge from NMM to NH_3 to DMEA (fig. 7). The influence of the variation of the pH-value is shown exemplary for the titanium dioxide paint in fig. 8. Here, a marked increase of negative surface charge can be observed for the paint having a higher pH-value (adjusted with amine during fabrication).

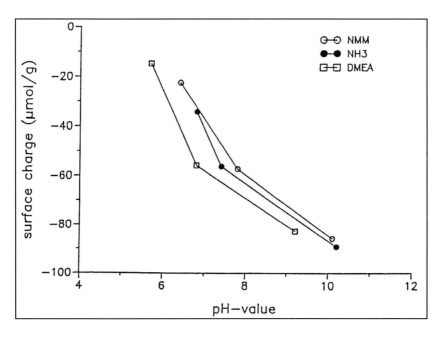

Fig. 7: Influence of amine variation on surface charge for carbon black paint

The results can be summarized as follows:

Carbon black and CuPc paints show for variation of amines some higher negative charge depending on the increasing base strength of the amine. For variation of the pH-value an increase of negative charge can be observed for higher pH-values. Due to that behaviour a better stability can be assumed, which will be confirmed by microscopy and Rub-out test [4].

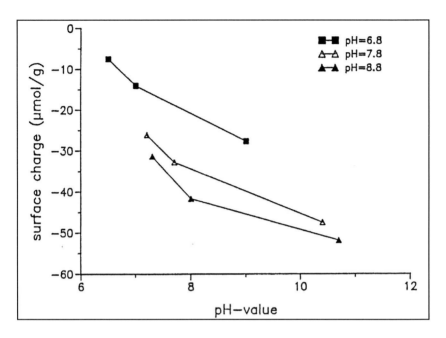

Fig. 8: Influence of ph variation on surface charge for titanium dioxide paint

5 Summary

Relating to water-based paints for industrial coatings, the importance of different system parameters, like type of amine and pH-value, were demonstrated. The effects on flow behaviour and surface charge were discussed. In this connection the most important physical measuring methods for characterizing disperse systems were presented using examples of practical interest.

References

01. G. Lagaly, in: Ullmann's Encyclopedia of Industrial Chemistry, Vol. A 7, VCH, Weinheim (1986) 341

02. H.-J. P. Adler, Farbe Lack , 97 (1991) 103

03. S. Cetin, Diplomarbeit, U-GH Paderborn (1993)

04. Th. Brock, 3rd Nürnberg Congress, Paper 17, (1995)

05. M. Breucker, Rheology, 93 (1993) 48

06. M. Breucker, Proc. XVI. Int. Conf. Org. Coat. Sci. Techn., Athens, (1990) 73

07. H.-J. P. Adler, Rheology, 92 (1992) 96

08. M. Osterhold, W. Schubert und W. Schlesing, Rheology, 92 (1992) 245

09. J. D. Ferry, Viscoelastic Properties of Polymers, Wiley & Sons, New York (1980)

10. A. A. Collyer, and D. W. Clegg (Eds.), Rheological Measurement, Elsevier Applied Science, London, New York, (1988)

11. H. A. Barnes and J. F. Hutton, An Introduction to Rheology, Elsevier, Amsterdam, (1989)

12. W.-M. Kulicke (Ed.), Fließverhalten von Stoffen und Stoffgemischen, Hüthig & Wepf, Heidelberg (1986)

13. R. Craft, Modern Paint and Coatings, (1991), No. 3, 38

14.   J. Schröder, Farbe Lack, 93 (1987) 715

15.   G. D. Parfitt, in: Dispersion of Powders in Liquids, G. D. Parfitt (Ed.) Applied Science Publishers, London, New Jersey (1981)

16.   L. Dulog and M. Hilt, Farbe Lack , 96 (1990) 180

17.   R. J. Hunter, Zeta Potential in Colloid Science, Academic Press, London (1981)

18   L. Dulog and M. Hilt, Farbe Lack ,95 (1989) 395

19.   J. P. Fischer and E. Nölken, Progr. Colloid & Polymer Sci., 77 (1988) 180

20.   J. Winkler, Farbe Lack ,97 (1991) 859

21.   J. Schröder, Farbe Lack ,97 (1991) 957

22.   J. P. Fischer and G. Löhr, in: Organ. Coat. Sci. & Technol. Vol. 8, G. D. Parfitt und A. V. Patsis (Eds.), Marcel Dekker lnc., New York (1986) 227

23.   M. Osterhold and K. Schimmelpfennig, Farbe Lack, 98 (1992) 841

24.   Firmenschrift Fa. Mütek, Herrsching, Deutschland

Chapter 5 – Particle Size Determination

Characterization of Disperse Systems (Part 2)

1 Introduction

Changes in particle size distribution caused, for example, by flocculation or variations of production conditions, can seriously affect important application properties, especially the rheological behaviour of the suspension or emulsion. This is why particle size analysis is used in quality control, to check that the product properties are constant, as well as to characterize dispersions during development. The particle analysers available on the market are based on very different physical principles. Laser instruments, electronic equipment and computers are being constantly improved and miniaturised, so that the use of instruments based on photon correlation spectroscopy (PCS) or laser diffraction is becoming more and more widespread. Whereas laser diffraction instruments are used mainly for measuring particles in the micrometer range, photon correlation spectroscopy, also known as dynamic light scattering or as quasielastic light scattering (QELS), is used for submicrometre particles.

2 Photon correlation spectroscopy (PCS)

In photon correlation spectroscopy, laser light is scattered by the individual particles of the sample which has to be diluted with water or organic solvents. The light scattered by the particles fluctuates because of Brownian motion and is detected by a photo-multiplier under a certain scatter angle and processed by a so-called correlator. A typical experimental set-up for PCS studies is shown in fig. 1 [1].

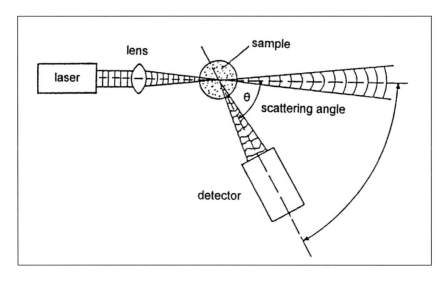

Abb. 1: Schematical set-up of photon correlation spectroscopy [1]

The fluctuating scattered light intensities contain the information about the distribution of particle sizes (or rather diffusion coefficients), which results from the detected light signal by using a special mathematical algorithm (auto-correlation) [2-4]. From the mathematical analysis of the correlation function and from the Stokes-Einstein relationship which links the diffusion coefficient with the hydrodynamic radius an intensity weighted distribution of particle sizes can be determined. According to Mie's theory a distribution of number or mass (volume) can be obtained from the intensity distribution [5]. The distribution of number gives the percentage of all particles having a certain particle size (diameter). The distribution of mass (volume) can be established from the distribution of number by weighting with the third power of the diameter.

Different averages are used for describing the particle size:

di: average intensity distribution
dn: average distribution of number
dm: average distribution of mass
dz: z-average-mean

Additionally, the so-called polydispersity is a measure of the width of a particle size distribution. In practice, the dn and dm values are normally used to characterize the particle size. All calculations are made under the assumption that the particles are spherical.

The present measurements were carried out with a Zetasizer 3 (Malvern), which can determine particle sizes in the range from approx. 3 nm to 3000 nm [1]. Prior to measurement all samples were diluted with deionised water. Especially in the case of broad or bimodal distributions the measurements must be carried out at different scattering angles (e. g. 60°, 90°, 120°).

For example, fig. 2 shows the intensity distributions of a mixture of 190 nm latices with 380 nm latices. The dm averages of this monomodal latices were determined by aerosol spectroscopy [6]. Here, the distributions are very different and a strong shift to higher particle sizes can be observed for a scattering angle of 60°. The two different particle size distributions (190 nm, 380 nm) can be easily identified in the distributions of number and mass (fig. 3, scattering angle 90°).

In the case of narrow monomodal particle distributions all measurements at different scattering angles yield nearly the same distributions and averages, as shown e. g. in fig. 4 (dn = 115 nm).

Figure 5 displays measurements for quality control of PUR dispersions. The "good" PUR-dispersion 1 has a narrow distribution at low particle diameters, while the "bad" PUR-dispersion 2 (out of tolerances) shows a bimodal size distribution with high particle sizes.

The PCS techniques can be easily used for different kinds of dispersions (binders), but PCS sometimes fails in the determination of particle sizes of pigments (because of aggregates/agglomerates). Methods like transmission electron microscopy (TEM), ultracentrifuge or field flow fractionation [7] are here more appropriate. Typical particle sizes of different substances are listed in table 1.

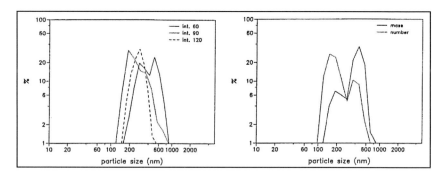

Fig. 2 (left): Intensity distributions of a bimodal sample at different scattering angles. Fig. 3 (right): Distribution of number and mass of fig. 2 at a scattering angle of 90°

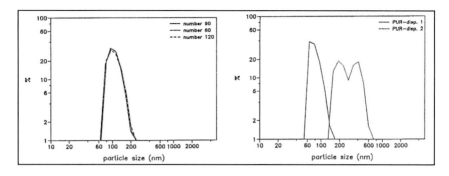

Fig. 4 (left): Distributions of number of a monomodal sample at differ-
ent scattering angles. Fig. 5 (right): Distributions of number of two
PUR-dispersions at scattering angle 90°

	particle size
resins	10 nm
PUR-dispersions	50-150 nm
titanium dioxide	200 nm
extenders (e. g. talc)	> 1 µm
aluminium-pigments	20 µm
powder coatings	20 µm

Table 1: Typical particle
sizes of various mate-
rials. The listed values
should be understood as
rough estimates of the
magnitude, not as abso-
lute values.

3 Laser diffraction

In laser diffraction instruments, the laser beam is diffracted by the particles under investigation. The diffraction pattern which is characteristic
for a specific particle size distribution is recorded by a detector in the
form of light energy distribution and then passed on to a computer
which calculates the particle size distribution from the diffraction pattern [8]. The analysed particles can be present in the form of liquid
suspensions or emulsions, whilst certain instruments can also handle
dry powders. The mathematical evaluation of particle size distribution

is based either on Fraunhofer's theory (> some micrometres) which does not assume knowledge of the refractive index of the particles, or on Mie's theory which should be used particularly for particles in the sub-micrometre range and, for very large particles, goes over into Fraunhofer's theory.

Laser diffraction instruments can be used to characterize e. g. aluminium-pigments, extenders (e. g. talc) or powder coatings. A typical particle size parameter calculated from the distribution is e. g. the median value (50 % value). The particle size distribution measurement of aluminium-pigments (fig. 6), which were dispersed in ethanol with an automatic sample dispersion unit, was performed with a Mastersizer X (Malvern) having an overall measuring range of approx. 0.1 µm to 600 µm [1]. The histogram represents the particular size classes, where all calculations are based on the assumption of spherical particles. Here, the median value is about 17 µm. Particle size distributions of materials can be investigated without a liquid medium by using a special dry powder feeder. Fig. 7 shows a corresponding particle size distribution of a powder coating having a median value of about 25 µm.

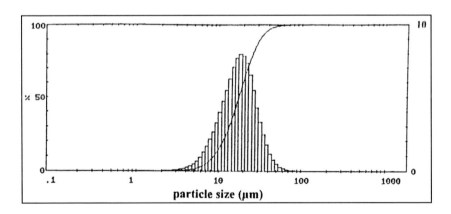

Fig. 6: Particle size distribution (volume distribution and sum curve) for an aluminium pigment in Fraunhofer approximation

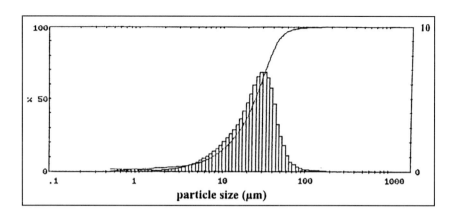

Fig. 7: Particle size distribution (volume distribution and sum curve) for a powder coating in Fraunhofer approximation

References

01.   Firmenschrift Fa. Malvern, Herrenberg, Deutschland

02.   J. Wagner, Chem.-lng.-Tech., 58 (1986) 578

03.   M. Weber, Chem.-lng.-Tech., 66 (1994) 707

04.   H. Z. Cummins and E. R. Pike (Eds.), Photon Correlation and Light Beating Spectroscopy, Plenum Press, New York (1974)

05.   P. Ludwig, Kontrolle, (1989), No. 4, 54

06.   J. P. Fischer, E. Nölken, Progr. Colloid & Polymer Sci., 77 (1988) 180

07.   T. Schauer, L. Dulog, Farbe Lack , 97 (1991) 765

08.   U. Vielhaber, Verfahrenstechnik, lnterkama-Report (1989)

Chapter 6 – Thermal Analysis

Dynamic Mechanical Analysis DMA

1 Introduction

The term "thermoanalysis" is applied to a number of different methods used to determine the physical and chemical properties of substances as a function of time and temperature. In thermoanalysis, the test specimens are subjected to a controlled temperature programme. "Thermoanalysis" essentially covers three methods of determination, based on the determination of changes in temperature, of mechanical properties and changes in weight.

2 Thermomechanical analysis

Thermomechanical analysis (TMA) is based on the following measuring principle: the test specimen is deformed by an applied static force, and the deformation is measured continuously as a function of temperature. This methods enables one to determine the glass transition temperatures and coefficients of expansion of solid substances such as polymers, not however to investigate their viscoelastic behaviour. Viscoelastic data can only be obtained by dynamic tests in which the reaction of the material in relation to a periodically changing force is measured. This is done by DMA (dynamic mechanical analysis).

2.1 Dynamic mechanical analysis

With the dynamic mechanical analysis (DMA) the mechanical material behaviour of polymers can be investigated. The film forming of liquid coatings is realized in different ways, depending on the ultimate coating property requirements:

- Creating of a felted molecular structure by the entanglement of high-molecular branched filamentary molecules.

- Constricting of the intermolecular mobility by incorporating of special molecular groups which activates the intermolecular forces.

- Forming three-dimensional networks due to chemical reactions of the reactive groups within the filamentary molecules.

These chemical and physical "cross-linking methods" influence the properties of the film, such as swellability, stone chip resistance, hardness, chemical resistance, so that a precise knowledge of the cross-linking is of particular importance.

2.1.1 Theoretical background

The theory of viscoelasticity describes the deformation behaviour of polymeric materials in all aggregate states under the constraint of small deformations and stresses. If a polymeric material is loaded with a sinusoidal force, its resulting deformation amplitude is out of phase with the force. The phase shift between force and deformation lies between ideally elastic (phase shift 0°) and an ideally viscous (phase shift 90°) behaviour. Elastic behaviour is the ability of a material to store deformation energy and is described with the storage modulus E'. The proportion of non-storable deformation energy is referred to as loss modulus E'' and is a measure of the dissipated deformation energy.

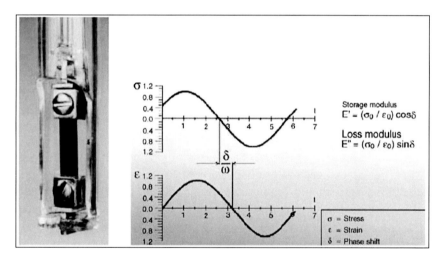

Fig. 1: Dynamic mechanical analysis in tensile mode

In dynamic mechanical analysis (DMA), the sample is loaded with a sinusoidal force (e. g. in the so-called tension mode) and the resulting deformation is measured over a wide temperature or frequency range (typical values: oscillation frequency 1 Hz, heating rate 10 K/min). In fig. 1 (based on [1]), the method is shown schematically. In this case, storage E' and loss modulus E'' are calculated from the ratio of the maximum values of stress σ_0 and strain ε_0 using the phase shift δ with the equations

$$E' = (\sigma_0 / \varepsilon_0) \cos \delta$$
$$E'' = (\sigma_0 / \varepsilon_0) \sin \delta$$

The loss factor is defined as the ratio of dissipated to stored energy

$$\tan \delta = E'' / E'$$

Fig. 2 (acc. [1]) shows a typical behaviour of a polymer system. In the brittle-elastic region (glassy state), the micro-Brownian motion is frozen. Binding distances and valence angles are elastically deformed under the influence of external forces. The storage module shows the highest value in this area. In the glass transition area begins the micro-Brownian motion. The storage module drops by several orders of magnitude of its value in the glassy state. The loss factor goes through a maximum due to distinct relaxation phenomena. The position of the maximum on the temperature scale can be used as a value for the glass transition temperature; as well as the maximum of the loss modulus. Both types of analysis can be found in the literature. The position of the glass transition on the temperature scale depends on the degree of cross-linking of a system. As cross-linking increases, the glass transition region in a system is shifted toward higher temperatures.

The glass transition region is followed by the more or less pronounced rubber-elastic region depending on the degree of cross-linking and entanglement. The value of the storage modulus in this range (E' min) is a measure of the crosslink density. In the flow region, the cross-linking points are dissolved, and the polymer system behaves increasingly viscous, i. e. it starts to flow.

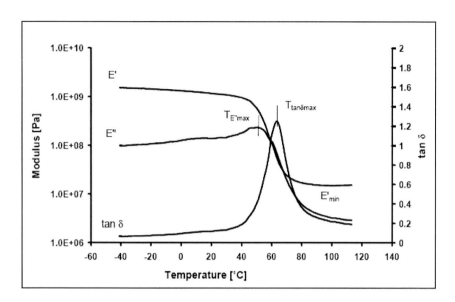

Fig. 2: Typical DMA results

2.1.2 Examples

When a polymeric system is converted from the uncross-linked to a cross-linked state by a chemical reaction, then the storage modulus increases in the course of the reaction, and the resulting network behaves rubber-elastically above the glassy state. Fig. 3 (acc. [2]) shows the temperature curves of the storage module of a nearly completely cross-linked coating film and an incompletely cross-linked coating film. The increase in the storage modulus with increasing temperature in the rubber-elastic range can be clearly observed.

The mechanical behaviour of ideal networks is described using the theory of rubber elasticity. The storage modulus in the rubber-elastic range is a measure of the cross-link density. In order to be able to measure in the uncross-linked state, a small coating strip or a fine coating saturated carrier-material is subjected to a sinusoidal tensile stress at a constant or rising temperature and the time or temperature profile of the storage module is determined.

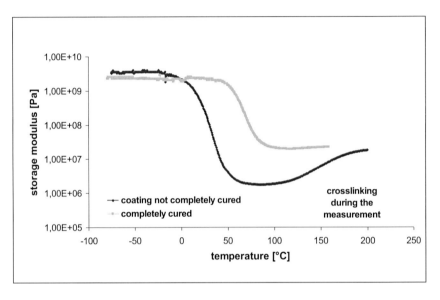

Fig. 3: Storage modulus as a function of temperature (comparison cross-linked and incompletely cross-linked system)

Fig. 4 (acc. [1]) shows the development of a clearcoat system as a function of time. The strong increase of the storage modulus from a time of about 7 minutes can be clearly seen, or at 140 ° C respectively (cross-linking). In this example, silk was used as the substrate material in this example.

In general, the crosslink density is particularly affected by the curing temperature and duration. Also important in this context are problems of under-curing (low crosslink density) or over-curing (possibly poor adhesion) and layer thickness variations.

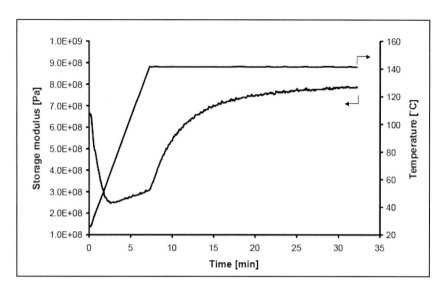

Fig. 4: Temperature profile and storage modulus of a silk/clearcoat system

The chapter on scratch resistance explicitly describes how crosslink density and scratch resistance relate to each other. Extensive presentations of dynamic mechanical analysis with many application examples can be found in [1, 3, 4].

3 Other methods

In differential thermoanalysis (DTA) the specimen and an inert reference standard are subjected to a temperature programme. The difference in temperature between the specimen and the reference standard, produced as a result of different thermal behaviour, is measured by means of a thermocouple. DTA may be used to qualitatively or semi-quantitatively characterize substances with regard to their endothermic (e. g. melting processes) and exothermic (exothermic reaction profiles) behaviour. Quantitative information (e. g. about reaction enthalpies) can be obtained by DSC (differential scanning calorimetry) instruments, with which one can measure differences in heat flow between the specimen and the reference standard. Here, two different measuring principles have become established – heat flow and power compensation DSC.

Thermogravimetric analysis (TG or TGA) records the change in weight of a substance during a specified temperature/time programme. This change in weight is normally determined with highly sensitive balances. The most important use of TG is to determine the decomposition temperature, and for content determination.

The combination of two different thermoanalytical techniques (TG and DSC) for the simultaneous examination of substances has been realized by a number of manufactures. These methods can also be coupled with other analytical methods such as mass spectrometry or FTIR.

References

01.   W. Schlesing, M. Buhk, M. Osterhold, Prog. Org. Coat.,
        49 (2004) 197

02.   M. Osterhold, Manuskript DFO-Seminar
        „Industrielle Lackiertechnik", (2002)

03.   W. Schlesing, M. Osterhold, H. Hustert, C. Flosbach,
        Farbe Lack, 101 (1995) 277

04.   W. Schlesing, Farbe Lack, 99 (1993) 918

Chapter 7 – Scratch Resistance

Methods for Characterizing the Scratch Resistance

1 Introduction

In addition to good levelling, high gloss and effect development the resistance to mechanical damage – by stone chippings and scratching – is particularly important in order to obtain a high-quality appearance for clearcoats. The brushes and dirt in a car wash, for example, produce scratches measuring only a few micrometres in width and up to several hundred nanometres in depth.

With this background several measuring methods have been discussed over the last 15 to 20 years in the automotive and paint industry. The objective was to obtain a clear characterization of the scratch/mar resistance of clearcoats. Procedures that create a single scratch have been developed and improved at the end of the 1990s (micro or nano scratch method) [1-8]. These methods are different from more practically oriented procedures that are based on relatively simple methods to try to test or even come close to reality (e. g. car wash brush method).

2 Characterization methods

2.1 Car wash brush methods (Amtec)

The clearcoat to be tested (applied on a standard metal sheet) is moved back and forth 10 times under a rotating car wash brush. The brush (PE) is sprayed with washing water during the cleaning procedure. Because the metal sheets used for the test are clean, a defined amount of quartz powder is added to the washing water as a replacement for street dirt (Amtec method). A gloss measurement e. g. in 20°-geometry is used to evaluate the scratch resistance. The initial gloss and the gloss after the cleaning procedure are measured. The percentage of residual gloss with regard to the initial gloss is a measure for scratch resistance. High values indicate good scratch resistance, the instrument is schematically shown in fig. 1 [8]. A commercial car wash brush instrument is manufactured by Amtec-Kistler/Germany.

This method was developed in a project working group of DFO (Deutsche Forschungsgesellschaft für Oberflächenbehandlung) two decades ago and has been specified in the standard DIN EN ISO 20566. More details can be found for example in [9].

2.2 Crockmeter

The crockmeter is used by large car manufacturers and represents a different type of strain. The crockmeter has been the standard device of the American Association of Textile Chemists and Colorists (AATCC). It has been mainly used to test textiles for colour fastness and abrasion.

This instrument (Atlas Materials Testing) is equipped with an electrical motor, so that a uniform stroke rate of 60 double strokes/min is reached. The sample is fixed on a flat pedestal. The sample is exposed to linear rubbing caused by a cylinder, other geometries are described in DIN 55654. A special testing material is attached to the bottom side of the cylinder, which is 16 mm in diameter; its downward force is 9 N. Ten double strokes are carried out over a length of 100 mm. For this method, the percentage of residual gloss is also used as a measure for scratching.

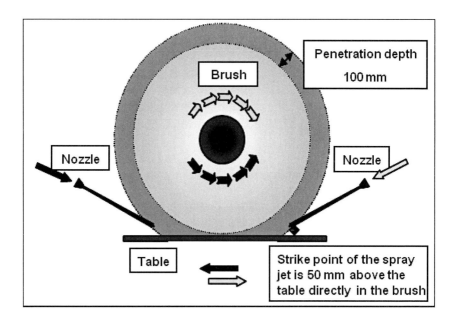

Fig. 1: Car wash brush method (Amtec)

In [8] a good correlation of both methods, crockmeter and laboratory car wash, in the range of good and bad scratch resistance can be observed. For the range of medium scratch resistance there is no clear correlation detectable. The reason for this could be the different degrees of strain in the two scratch test methods. The Amtec test is the one with a rather higher strain load.

2.3 Rota-Hub method

The Rota-Hub-Scratch-Tester has been developed by Bayer AG approx. 15 years ago. In this test, a carriage moves in x- and y-directions. A rotating disc with the scratching medium (e. g. paper) is attached to the carriage. The rotating disc is lowered onto the sample. This way the sample is strained by a rotating disc that also moves in x- and y-directions. Feed rate, rotation velocity, and contact pressure can be randomly selected and have been optimized.

For the automotive industry copy paper has proved to be a suitable scratching material because it causes scratches similar to those caused by a car wash. The resulting damage is a meander-shaped scratch pattern that can be measured using the parameters gloss and haze [10].

2.4 Micro-scratch experiments

In micro-scratch experiments single scratches of characteristic and realistic phenotype can be generated. The indentation depth depends on the applied force and the indentation body (indentor). Indentation depth is usually in the range of 1 µm and smaller. A diamond indentor (radius at the peak 1 – 2 µm) is pushed onto the sample surface applying a defined force to generate the scratch while the sample is moved in a linear direction at constant velocity underneath the indentor (see fig. 2). The applied normal force can be constant or progressive during the scratching procedure.

Basically it is possible to measure the tangential forces and the indentation depth during the scratching procedure. Deformations or damages can be observed with a microscope or additionally with a video camera. The profile of the generated scratches can be measured using AFM technology. In combination with the parameters measured during the scratching procedure it is possible to calculate physical parameters which allow conclusions about the elastic and plastic deformation behaviour and the fracture behaviour.

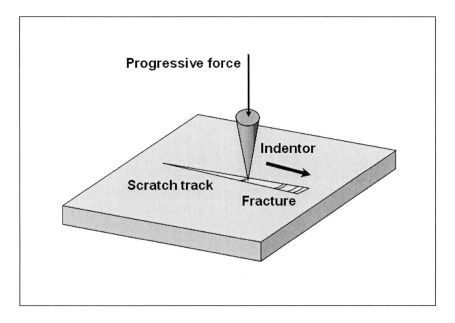

Fig. 2: The principle of single-scratch tests

2.4.1 Tests with progressive load

In this context a method was developed in the DuPont Marshall Lab. (Philadelphia) that generates and evaluates a single scratch on a surface (nano-scratch method). This method has been used for tests along with development work for clearcoats.

A commercial instrument by the Swiss company CSM (now part of Anton Paar company) was tested in the course of a project at the Research Institute for Pigments and Paints (Forschungsinstitut für Pigmente und Lacke e.V. (FPL), now part of IPA/Stuttgart) around the year 2000 and later. This instrument is based on the method developed at the DuPont Marshall Lab. Objective of several projects – where manufacturers of paints, raw materials and automobiles worked together – was the evaluation of the nano-scratch-tester regarding reproducibility, accuracy and applicability under the aspect of a realistic determination of scratch resistance.

The key point in this method is the determination of the critical load where first irreversible cracks or fractures are generated and which therefore indicates the transition from plastic deformation to significant/lasting damages. For that, the normal load is constantly increased and indentation depth and tangential load are simultaneously recorded. The transition from plastic deformation to the fracture range is indicated e. g. by unsteadiness or fluctuations in the detected load flow and the indentation depth. As a typical dimension besides the critical load it is common to determine the residual indentation depth after scratching – typically at a normal load of 5 mN – in the range of plastic deformation.

In figure 3 the transition range of an acrylate/melamine system is demonstrated. The top picture shows an AFM image of the scratch ridge itself where cracks are clearly visible. The bottom picture is taken with a microscope at a magnification factor of app. 1000, see also [8, 11, 12].

If values from nano-scratch tests are compared to results from car wash tests (see fig. 4 [8]), a certain correlation to the critical load can be observed. From this it can be concluded that the Amtec test rather causes a significant damage than strains in the plastic range [13].

3 Dynamic mechanical analysis

For certain clearcoat systems a partial healing of scratches can be observed on the time scale. In literature this is known as the reflow effect [14]. Thermal relaxation phenomena may be used for a physical explanation of this effect. In connection with scratch/mar resistance the cross-linking density of clearcoats is also a decisive factor. Dynamic mechanical analysis (DMA) has been established as a method to determine cross-linking density [14-16].

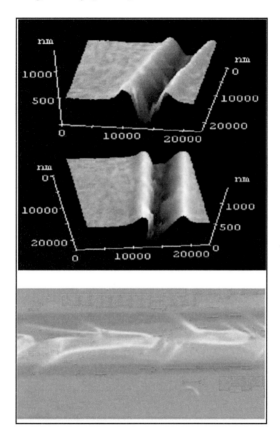

Fig. 3: AFM image (top) and microscopical (light microscope) image (bottom) of cracks in an acrylate/melamine system

Several ranges can be characterized when observing the behaviour of the storage modulus E´ as a function of temperature. The glass transition range is followed by the rubber-elastic range, which can be more or less distinctly depending on the degree of cross-linking. In this range the value of the storage module E´ is connected with the cross-linking density. The value of E´ at the local minimum is often chosen as a quantitative measure for cross-linking density.

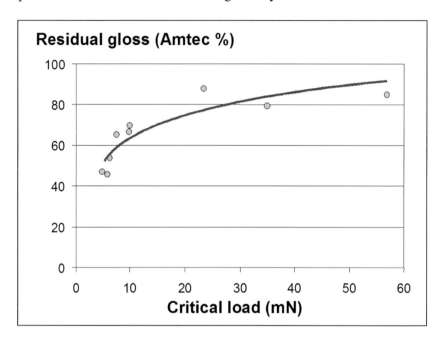

Fig. 4: Comparison of values from Amtec measurements to values for critical load from nano-scratch tests

The clearcoats were examined by submitting free films to a tension test. A DMA 7 instrument by Perkin Elmer was used for these tests. In this context, measurements of coating systems are described e. g. in [16].

Fig. 5 shows a comparison of the results of DMA analysis and values obtained in car wash simulations [8].

A clear correlation between E´ as a measure for cross-linking density and the value for scratching can be observed. The quality of the scratching level (high residual gloss) increases with increasing cross-linking density [17].

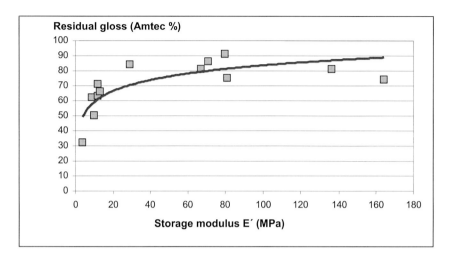

Fig. 5: Correlation of scratch resistance (Amtec values) and cross-linking density (storage modulus E´)

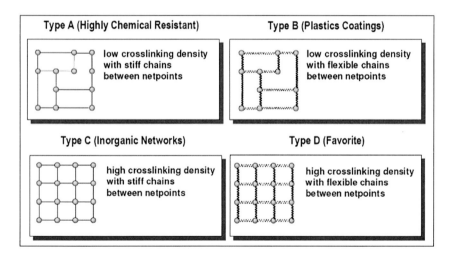

Fig. 6: Different network types

Fig. 6 shows a short comparison of different network types and brief remarks of their properties [18, 19, 20]; with type D as a favorite for automotive clearcoats.

4 Influence of weathering

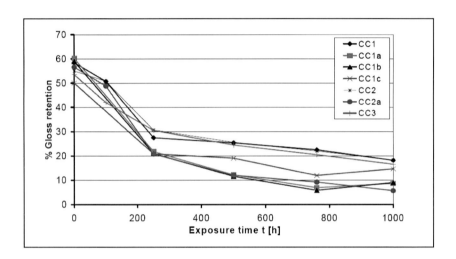

Fig. 7: Gloss retention (Amtec) troughout SAEJ1960 weathering

The influence of weathering on scratch resistance is demonstrated in fig. 7, showing the gloss retention after Amtec testing of various clear-coats as a function of exposure time of SAEJ1960 weathering. All samples that were investigated show a significant drop of residual gloss after a short exposure time. A clear differentiation between the clearcoats with (CC1, CC2, CC3, CC1c) and without light-screener (CC1a, CC1b, CC2a) is noticeable. Afterwards the residual gloss decreases slightly [21, 22]. Other studies can be found e. g. in [23].

5 Summary

In this chapter, several methods to determine the mar (scratch) resistance of clearcoats are presented. Methods that create a single scratch are different from more practically oriented procedures that are based on relatively simple methods to try to test or even come close to reality (e. g. car wash brush method). The methods were reviewed briefly considering also other physical properties like cross-linking density or weathering.

Recent investigations show that scratch/mar resistance is always an important factor to characterize physical properties of coating surfaces, for example in analysing automotive clearcoats containing silane modified blocked isocyanates [24]. Their results showed that a close correlation existed between the scratch resistance data obtained from car-wash and nano-scratch tests for certain clearcoats.

Also, scratch resistance of exterior clearcoats and polycarbonate hardcoats were examined and discussed recently by [25].

Studying the literature there is often no clear separation between the expressions mar and scratch resistance. Mar should refer to light surface damages encountered in the real field that are usually shallow and narrow while scratch refers to medium or more severe damages [26]. However, several authors handle this uncertainty in writing scratch/mar.

References

01.   G. S. Blackmann, L. Lin, R. R. Matheson, ACS Symposium Series: Microstructure and Tribology of Polymer Surfaces (1999)

02.   L. Lin, G. S. Blackman, R. R. Matheson, Prog. Org. Coat., 40 (2000) 85

03. K. Adamsons, G. Blackmann, B. Gregorovich, L. Lin, R. Matheson, Prog. Org. Coat., 34 (1998) 64

04. J. L. Courter, Proc. 5th Nürnberg Congress, Nuremberg, Germany, (1999) 351

05. E. Klinke, C. D. Eisenbach, Proc. 6th Nürnberg Congress, Nuremberg, Germany, (2001) 249

06. L. Lin, G. S. Blackman, R. R. Matheson, Materials Science and Engineering, A317 (2001) 163

07. G. Wagner, M. Osterhold, Mat.-wiss. u. Werkstofftech., 30 (1999) 617

08. M. Osterhold, G. Wagner, Prog. Org. Coat., 45 (2002) 365

09. E. Fischle, Farbe Lack, (2009), No. 04

10. E. Klinke, M. Kordisch, C.D. Eisenbach, T. Klimmasch, Farbe Lack, 108 (2002), No. 2, 55

11. M. Osterhold, Europ. Coat. Journal, (2005), No. 09, 34

12. M. Osterhold, Progr. Colloid Polym. Sci., 132 (2006) 41

13. K.-F. Dössel, SURCAR, Cannes, France, 2001

14. R. Gräwe, W. Schlesing, M. Osterhold, C. Flosbach, H.-J. Adler, Europ. Coat. Journal, (1999), No. 3, 80

15. W. Schlesing, Farbe Lack, 99 (1993) 918

16. W. Schlesing, M. Buhk, M. Osterhold, Prog. Org. Coat., 49 (2004) 197

17. E. Frigge, Farbe Lack, 106 (2000), No. 7, 78

18. M. Osterhold, Vortragstagung der GDCh-Fachgruppe Anstrich-stoffe und Pigmente, Eisenach, Germany, 2005

19. C. Flosbach, XXVIII Fatipec Congress, Paper II.C-1, Budapest, Hungary (2006)

20. K.-F. Dössel, in: Automotive Paints and Coatings, 2nd Ed., H.-J. Streitberger, K.-F. Dössel (Eds.), Wiley-VCH (2008)

21. M. Osterhold, B. Bannert, W. Schubert, T. Brock, Macromol. Symp. 187, (2002) 823

22. B. Bannert, M. Osterhold, W. Schubert, T. Brock, Europ. Coat. Journal, (2001), No. 11, 30

23. C. Seubert, M. Nichols, K. Henderson, M. Mechtes, T. Klimmasch, T. Pohl, J. Coat. Technol. Res., 7 (2010), 159

24. S. M. Noh, J. W. Lee, J. H. Nam, J. M. Park, H. W. Jung, Prog. Org. Coat., 74 (2012) 192

25. C. Seubert, K. Nietering, M. Nichols, R. Wykoff, S. Bollin, Coatings, 2 (2012) 221

26. Z. Ranjbar, S. Rastegar, Prog. Org. Coat., 64 (2009) 387

Chapter 8 – Surface Structure

Analysis of the Surface Structure of Substrates and Coatings

1 Introduction

In addition to colour, effect and gloss, the visual impression of a painted surface is influenced especially by the surface structure (levelling, waviness, orange peel, appearance). To characterize the structure of a surface, mainly two different measuring methods – profilometry and 'wave-scan' – have been established in the automotive and coatings industry.

The mechanical profilometry combined with Fourier techniques (FFT) yields detailed information of the surface topography, and substrate influences or other effects on the final coating appearance can be described [1-18]. To simulate the visual impression obtained from optical inspection of surface structures, the German company Byk-Gardner developed the so-called 'wave-scan' instruments. In addition to measurements on glossy surfaces, i. e. topcoats/clearcoats, the 'wave-scan *dual*' allows to detect the appearance of surfaces with lower gloss (medium glossy), e. g. primer-surfacers and sometimes EC.

Basic relations and relevant application examples from the areas 'metal, plastics, coating', investigated with profilometry and 'wave-scan' in the course of time, will be summarized in this chapter.

2 Methods

2.1 Mechanical surface characterization

Surface profiles presented in this chapter were measured by mechanical profilometry using the Hommeltester T 8000 (Hommel-Etamic, Germany). For all measurements, a dual-skid tracing system or a so called datum system (without skids) with a diamond tip radius of 5 µm was used. The vertical resolution of this mechanical profilometric system is approx. 0.01 µm. The surface profiles were recorded over a scan length of 48 or 15 mm. A cut off wavelength of 8 mm for a scan length of 48 mm was used to separate between roughness and waviness profile.

The evaluation of the mechanical profile measurement according to typical roughness parameters – e. g. average roughness Ra – gives an integrated information about the surface structure. In comparison to roughness parameters, Fourier techniques (FFT) yield a more detailed characterization of the surface structure.

In fig. 1 typical surface profiles from substrate to topcoat are shown.

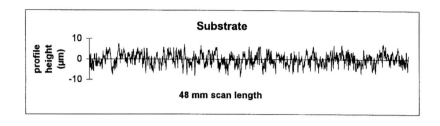

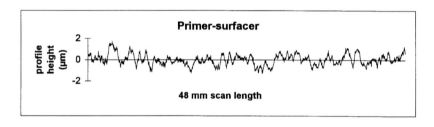

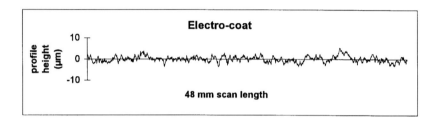

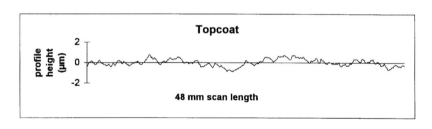

Fig. 1: Typical surface profiles from substrate to topcoat

When using Fourier analysis, the curve shape of the unfiltered profile data is separated into a sum of sine and cosine waves with different amplitudes and wavelengths.

In a so-called autopower spectrum, the intensity (square of amplitudes) of the calculated sine/cosine waves representing the surface profile is plotted versus the corresponding wave number (reciprocal wavelength).

To illustrate this procedure, fig. 2 shows a typical surface profile and the corresponding autopower spectrum after Fourier analysis for wave numbers > 0.1 mm^{-1} (wavelengths < 10 mm).

This shows that the spectrum is especially dominated by high intensities in the wave number range from 0.1 to 1 mm^{-1} (wavelengths from 10 to 1 mm).

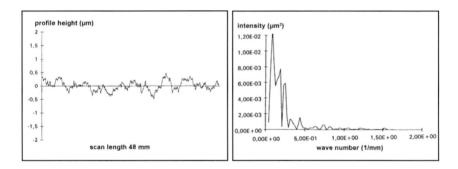

Fig. 2 : Typical surface profile of a topcoat (left) and the corresponding autopower spectrum (right)

For wavelengths from 10 to 1 mm (integral 1, long waviness) and from 1 to 0.1 mm (integral 2, short waviness), the intensities of the autopower spectra are added up and used for further evaluation of the surface structures (see e. g. [14]).

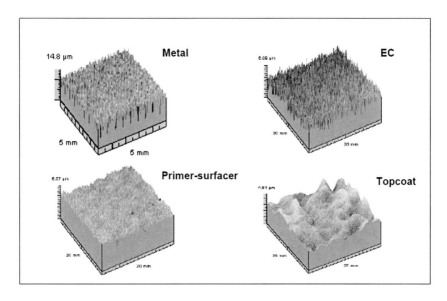

Fig. 3: Topographies of metal substrate and coating layers

In fig. 3 typical surface topographies from substrate to topcoat are shown in a pseudo-3-dimensional presentation. The sample is moved by a precise positioning table for a small distance between two line scans. In general, a decrease of the amplitudes and a change in wave-lengths can be observed.

2.2 Optical surface characterization

2.2.1 wave-scan

The optical determination of coating structures was carried out by the wave-scan *dual* or with the former instrument wave-scan DOI. Here, the measuring principle is based on the modulation of the light of a small laser diode reflected by the surface structures of the sample. The laser light shines on the surface at an angle of 60°, and the reflected light is detected at the gloss angle (60° opposite). During the measurement, the 'wave-scan' is moved across the sample surface over a scan length of approx. 10 cm. The signal is divided into 5 wavelength ranges in the range of 0.1 to 30 mm and processed by mathematical filtering. For each of the 5 ranges a characteristic value (Wa 0.1-0.3 mm, Wb 0.3-1.0 mm, Wc 1.0-3.0 mm, Wd 3.0-10 mm, We 10-30 mm) as well as the typical wave-scan-values longwave (LW, approx. 1-10 mm) and shortwave (SW, approx. 0.3-1 mm) are calculated. Low wave-scan-values mean a smooth surface structure. Additionally a LED light source is installed and illuminates the surface under 20° after passing an aperture. The scattered light is detected and a so-called dullness value (du, < 0.1 mm) is measured. A change to an IR-SLED for the low gloss area allows to measure samples with medium glossy surfaces.

3 Metal substrates

In this part an example of former studies on steel substrates [10] is presented. Three different phosphated steel surfaces with different roughness levels were evaluated. The samples were chosen in order to cover the range of roughness from smooth to rough substrates (Ra 0.8 to 2.3 μm) supplied for automotive body sheets, where – typically – sheets with an average roughness between Ra = 0.8 μm (smooth) and Ra = 1.6 μm (middle, upper limit) were used. All samples were coated with cathodic electrodeposition paint, primer-surfacer and automotive topcoat with typical layer thicknesses. One set of painted substrates was baked in horizontal position, a second set vertically.

As shown in fig. 4, the integrated values increase as a function of substrate surface roughness and baking position. For the vertical baking position, a strong enhancement of the long waves can be observed, while the short wave values hardly increase compared to the samples baked horizontally. The short wave values depend on substrate conditions (roughness), not strongly on baking position.

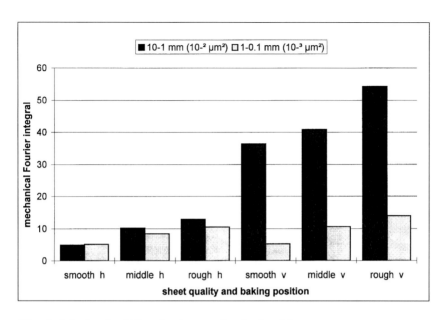

Fig. 4: Mechanical Fourier integrals obtained by measurements on topcoats

4 Investigations on plastics coatings

Investigations on plastics coatings are of special interest to achieve a similar structure development on a painted car not depending on the different substrate materials (steel or plastics) used. Against this background parameters influencing the surface of plastics parts are becoming recently more important.

4.1 Measurements on reinforced plastics (structure effect)

Polymeric substrates with different amount of glass fibres were painted with a typical coating system for plastics. The complete coatings were measured by wave-scan and mechanical profilometry [12]. Optical and mechanical measurements (Ra) show both the same trend, an increasing measuring value of structure for an increasing amount of glass fibre. In addition to the effect of the amount of fiber content also a dependence on the type of the used fibre can be observed in a different study [13].

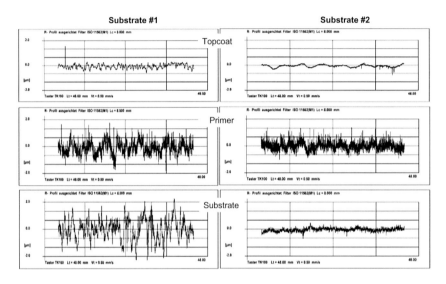

Fig. 5: Comparison of two substrates and coating layers (mech. pro-filometry, roughness profiles, scan length 48 mm, y-scale +- 2 μm)

This effect is also demonstrated in fig. 5 showing a comparison of two substrates with high and low surface structure. The amplitude of the structures is reduced, but the basic structure of the substrate is partially transferred through each single coating layer and influences finally the topcoat appearance.These results are very important due to the effect, that the substrate structure clearly influences the final coating structure. Based on surface structure measurements a pre-selection of plastics substrates can be done to obtain an optimal structure (smoothness) of the coating system applied [15].

5 Investigations on medium glossy surfaces

Several test sets have been measured to investigate the effects of coating variation (EC and primer-surfacer) and substrate topographies on the different coating layers. These studies are described in detail in [18]. The investigations and main results for varying the steel substrate will be summerized in this chapter.

5.1 Variation of substrate

Most of the 28 studied substrates were in the range which is regularly used in the automotive industry for outer car body parts (Ra 0.9-1.5 μm, RPc > 60). In fig. 6 the profilograms of substrates and the corresponding EC (electro coating) surfaces are demonstrated by 4 selected samples.

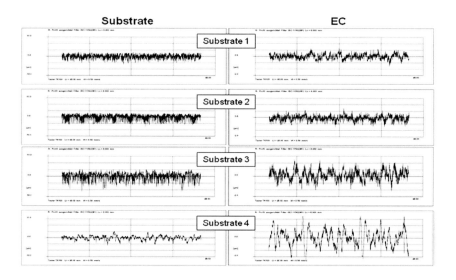

Fig. 6: Roughness profile of panel and EC surfaces (variation substrate). Scan length 48 mm, y-scale +- 10 μm (substrate) +- 2 μm (EC)

The Ra values of the substrates 1-3 were in a roughness range from 0.9-1.5 µm, with peak counts of 60-75 points/cm. Substrate 4 shows a good Ra value of 0.8 µm, but a very low number of 12 peaks/cm. The painting of this substrate is more difficult because of the low peak number and would not be used for the outer shell of a car body. The EC application as well as the whole following coating process was carried out with the same material and under identical application conditions.

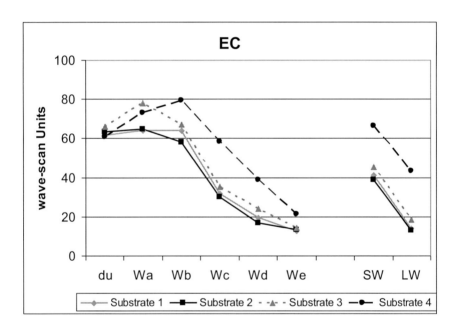

Fig. 7: Structure spectra, SW and LW of EC (variation substrate)

The measurement of the EC (fig. 7) with the wave-scan dual could be carried out without any difficulty and shows for the 4 example substrates comparable ranking of EC and topcoat structure (fig. 8).

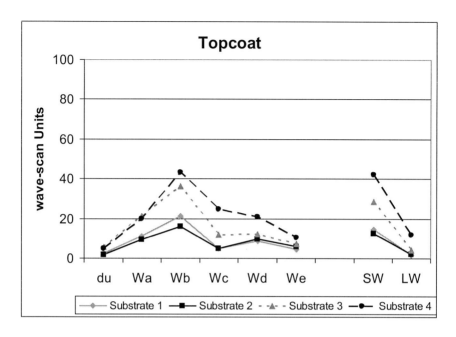

Fig. 8: Structure spectra, SW and LW of topcoats (variation substrate), horizontal application

Considering the whole sample set, the linear correlation coefficient of profilometric structure measurement between substrate and EC structure is r≈0.9 for the long wave range (integral 1) and r≈0.7 for the short wave range (integral 2). There is a correlation coefficient of 0.8 between EC and clearcoat structure for horizontal application [18].

6 Further studies and summary

Further studies considering the influence of sheet-type, substrate roughness, deformation and paint system on the appearance of the painted surface were summarized in [15]. Additional investigations by variation of coating systems with lower gloss (EC, primer-surfacer) can be found in [18].

Influences of the forming/deep-drawing process yielding to an increased waviness of metal substrates were reported in [19].

Investigations of the mechanisms for the formation of surface structures on coating layers are considering for example the flow behaviour of the liquid paint driven by viscosity, surface tension and other parameters. Modeling and simulations of the interrelations of intrinsinc physical and process parameters are used to understand the causes of coating film appearance [20-22].

The mechanical profilometry can be applied for all substrates and coating layers. The use of the wave-scan dual is limited by dullness and a distinct surface structure. The optical investigation of EC is possible, if the surface structure of the used substrate in combination with the self structure of the coating material is not too distinguished. In the daily work primer-surfacers can usually be investigated without reservation.

References

01.   D. W. Boyd, Proc. XIII. Int. Conf. Org. Coat. Sci. Techn., Athens, (1987) 59

02.   F. Fister, N. Dingerdissen, C. Hartmann, Proc. XIII. Int. Conf. Org. Coat. Sci. Techn., Athens, (1987) 113

03.   K. Armbruster, M. Breucker, Farbe Lack, 95 (1989) 896

04.   T. Nakajima, Y. Yoshida, Y. Miyoshi, T. Azami, Proc. XVII. Int. Conf. Org. Coat. Sci. Techn., Athens (1991) 227

05.   W. Geier, M. Osterhold, J. Timm, Metalloberfläche, 47 (1993) 30

06.   A. F. Bastawros, J. G. Speer, G. Zerafa, R. P. Krupitzer, SAE Technical Paper Series 930032

07.   J. Timm, K. Armbruster, M. Osterhold, W. Hotz, Bänder, Bleche, Rohre, 35 (1994), No. 9, 110

08.   M. Osterhold, W. Hotz, J. Timm, K. Armbruster, Bänder, Bleche, Rohre, 35 (1994), No. 10, 44

09.   M. Osterhold, Prog. Org. Coat., 27 (1996) 195

10.   M. Osterhold, Mat.-wiss. u. Werkstofftech., 29 (1999) 131

11.   O. Deutscher, K. Armbruster, Proc. 3rd Stahl-Symposium, Düsseldorf, Germany (2003)

12.   H. Stegen, M. Buhk, K. Armbruster, Paper TAW Seminar „Kunststofflackierung – Schwerpunkt Automobilindustrie", Wuppertal, Germany (2002)

13.   K. Armbruster, H. Stegen, Proc. DFO Congress „Kunststofflackierung", Aachen, Germany (2004) 78

14.   M. Osterhold, K. Armbruster, Proc. DFO Congress „Qualitätstage 2005", Berlin, Germany (2005) 4

15.   M. Osterhold, K. Armbruster, Prog. Org. Coat., 57 (2006) 165

16.   O. Deutscher, BFI, Düsseldorf, Carsteel-Bericht (2008)

17.   M. Osterhold, K. Armbruster, Proc. DFO Congress „Qualitätstage 2008", Fürth, Germany (2008) 7

18.   M. Osterhold, K. Armbruster, Prog. Org. Coat., 65 (2009) 440

19.   D. Weissberg, Proc. DFO Congress „21. Automobil-Tagung 2014", Augsburg, Germany (2014)

20.   C. Hager, M. Schneider, U. Strohbeck, Proc. ETCC 2012, Lausanne, Switzerland (2012)

21. M. Hilt, M. Schneider, „Die Entstehung von Lackfilmstruktu-
     ren verstehen", www.besserlackieren.de, 07.03.2014

22. O. Tiedje, Proc. DFO Congress „21. Automobil-Tagung 2014",
     Augsburg, Germany (2014)

Chapter 9 – Surface Tension

Surface Tension and Physical Coating Properties

1 Introduction

In order to achieve a good wetting of the substrate with the liquid paint, the surface tension of the substrate must be as high as possible. Contamination of the surface with substances of low surface tension can cause wetting and adhesion problems. To determine the surface tension of the solid surface, the contact angle of several test liquids of known surface tension must be measured.

2 Theoretical background

In the literature, the basic equation used to describe the interaction of the surface tensions of a liquid and a solid is called the Young equation [1]:

$$\sigma_s = \sigma_{sl} + \sigma_l \cos \theta$$

with σ_s and σ_l representing the surface tensions of the solid and liquid phases and σ_{sl} representing the interfacial tension between the solid and the liquid. The contact angle θ represents the angle that forms when the liquid comes into contact with the solid (fig. 1).

According to a method of Zisman [2], liquids of different surface tension are applied to the solid surface and the resulting contact angles are measured. After plotting the cosine of the determined contact angle and the surface tension of the used test liquids it can be extrapolated to $\cos \theta = 1$ (corresponds to $\theta = 0$, just complete wetting). The value obtained by extrapolation represents the critical surface tension of wetting. In the more far-reaching theoretical approaches which deal with the problems of interfaces, the surface tension of each phase is separated into components considering only the dispersion forces on the one hand and those components comprising all polar interactions on the other:

$$\sigma = \sigma^d + \sigma^p$$

with σ^d representing the disperse component and σ^p representing the polar component. On the basis of this approach, Owens and Wendt [3] and Kaelble [4, 5] developed a relation linking the contact angle with the polar and disperse components of the surface tensions of liquid and solid. After conversion of this equation to a general linear equation, a simple graphic analysis after measurement of the contact angles of various test liquids enables the determination of the solid surface tension (surface free energy) (see also chap. 3.2.2). Table 1 lists some typical surface tensions of liquids and solids.

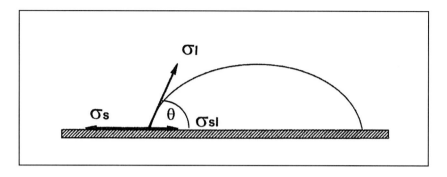

Fig. 1: Definition of the contact angle with vector representation of the surface tensions

3 Procedures for evaluating surface tensions

3.1 Surface tension of liquids

The ring method according to Noüy [6] is a widely used, relatively simple method of measuring the surface tension of liquids. It measures the maximum force F necessary to pull a horizontally suspended ring out of the liquid surface (figure 2). The surface tension as force per length is given by the equation $\sigma = F/4\pi r$. The ring usually consists of a maximum of 0.4 mm thick platinum or platinum-iridium wire and has an average diameter (2r) of about 19 mm. Instead of the ring a frame can also be used. Detailed descriptions of these methods with correction rules for the evaluation of results can be found in several DIN standards [7, 8].

The plate method according to Wilhelmy is also based on a force measurement as the ring method, but has the advantage that no hydrostatic corrections must be performed. In this method, the liquid surface is brought into contact with the lower edge of the plate and the force resulting from the wetting of the plate is measured [7]. For other, in practical applications minor used methods for determining the surface tension of liquids (e. g. capillary method, droplet size method, etc.), see [9].

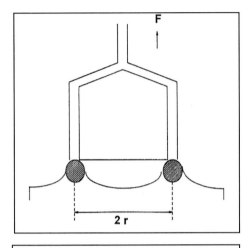

Fig. 2: Ring method for determing surface tension

	σ total (mN/m)	σ polar (mN/m)
Liquids		
Water	72,8	51,0
Formamide	58,2	18,7
Glycerol	63,4	35,1
Ethylene glycol	47,7	18,2
Ethanol	22,1	4,6
Plastics		
Teflon	18,5	0
Polypropylene		
(PP-EPDM)	21-23	2-3
R-TPU	31-32	12-20
R-RIM	30	14
Various		
Galvanized steel		
(phosphated)	36-49	27-39
Glas	65	59
Coating (liquid)	27-35	

Tab. 1: Surface tensions of some liquids and solids

3.2 Surface tension of solids

3.2.1 Union Carbide test method / test inks

The simplest method for surface tension measurement of solids is the Union Carbide test (UC test) [10]. The method is based on the premise that a liquid having a lower surface tension than that of the solid completely spreads on the solid. In practice, therefore, a test liquid of low surface tension is used and a thin liquid film is applied to the sample by means of a brush.

If the liquid film holds together for more than two seconds and does not break apart into droplets, the test liquid with the next higher surface tension is selected from a uniformly graduated series. This process is repeated until a liquid is reached whose film breaks up or contracts after two seconds. The surface tension value of the liquid that has just wetted is then assigned to the sample. A comparison of this method with the contact angle method (chapter 3.2.2) and the shortcomings of the test ink method will be discussed later.

3.2.2 Contact angle method

For the exact evaluation of the surface of a solid (e. g. after pretreatment) or the acting surface forces it is important to divide the total surface tension into its polar and disperse elements. In order to make this division, the contact angles formed on a solid surface by different test liquids of known surface tensions (polar, disperse elements) are measured [5].

The very time-consuming manual measurements of contact angles can be carried out within an acceptable time scale using automatic contact angles measuring instruments.

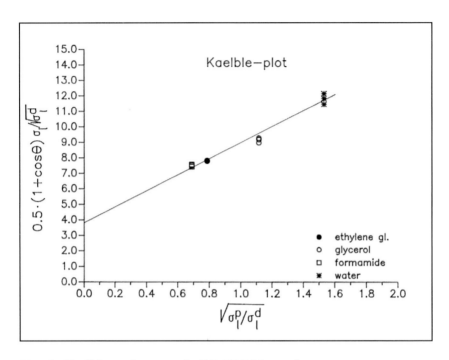

Fig. 3: Kaelble evaluation of a PP-EPDM panel

A fully automatic device (AURAM) was developed at the end of the 1980s in cooperation between the engineering department of Hoechst AG and the central physics department (central research department) of Herberts GmbH [11]. Under video screen control, the droplet of the test liquid is deposited vibration-free onto the specimen surface. With the aid of a zoom lens, the relevant droplet area is imaged and analysed. From the contours of the droplet, the contact angle can be calculated. Various computer-aided calculations can be carried out directly following the measurement.

In a so-called Kaelble-plot, values, essentially determined by the measured contact angles, are plotted against the square root of the ratio of the polar to the disperse element of the surface tension of each test liquid. Using these values, a regression line is plotted. Its slope gives the

102

polar, and the ordinate intercept the disperse element of the solid surface tension (fig. 3). In this example, the test liquids were double-distilled water, formamide, glycerol and ethylene glycol. All of the surface tension measurements presented below were obtained using this Kaelble method. Meanwhile the contact angle method is also described in detail in a series of standards (DIN 55660). The test liquids can vary.

4 Examples

4.1 Results at pretreated plastics

In the coatings industry, many different problems can arise in painting plastic parts in terms of the wettability of the plastics surface and adhesion of a coating to it. These difficulties in wetting and adhesion are due to the fact that many plastics, especially the polyolefins, are inherently of low surface tension, but good surface wetting and adhesion require large surface tension values. For good adhesion of the coating on the substrate surface, in particular the polar component of the surface tension is of great importance [12, 13].

Due to this fact, in industry, e. g. for bumpers, a variety of pretreatment processes (flaming, plasma, etc.) were used with the aim of increasing the surface tension of the plastics [14-20]. Similar problems occur, for example in the printing industry, which uses the corona process to improve the wetting and adhesion of the ink on polyethylene films [21, 22]. The various pretreatment methods activate the surface and incorporate oxygen. In this way, different functional groups, such as hydroxyl, carbonyl and carboxyl groups arise, which contribute to increase the surface energy, especially the polar part [20, 23-26].

The surface tension values of untreated PP-EPDM materials (elastified polypropylene) are between 21 and 23 mN/m (total) with a low polar component of 2 to 3 mN/m. Surface tension measurements of flame-treated PP-EPDM plates were carried out using the AURAM automatic contact angle measuring device. After flame treatment, the total surface tension increases and the polar component is strongly enhanced. After the third flame treatment procedure, surface tension and polarity no longer change (fig. 4).

Furthermore, plasma-treated (microwave discharge in oxygen) and sulfonated PP-EPDM samples were also investigated. It was found that the surface tension increases strongly by all three pretreatment methods considered and changes little over a longer period of time [34].

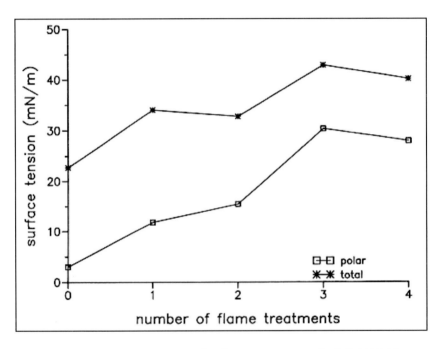

Fig. 4: Total surface tension and polar component of a PP-EPDM sample for different numbers of flame treatments

The increase of the surface tension is due almost exclusively to the increase of the polar component of the surface tension. This increase of the polar component can be explained by the integration of polar groups into the surface. This increase in the polar fraction can be explained by the incorporation of polar groups in the surface, which resulted in distinct bands in recorded IR spectra.

As a further diagnostic tool, ESCA measurements were taken of flamed samples. These investigations should support the thesis that the increase of the polar part of the surface tension is caused by oxidation of the surface carbon of the samples.

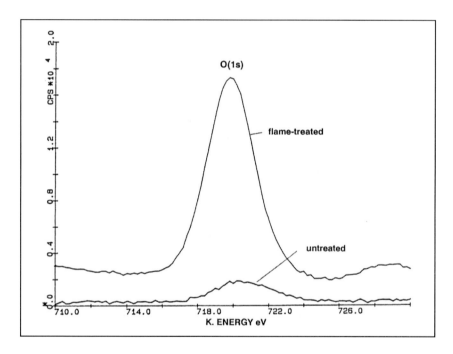

Fig. 5: ESCA oxygen spectra of a flame-treated and untreated sample

The C(1s) signal from aliphatic hydrocarbons, which has a binding energy of 284.6 eV, is used for the analysis of the bonding conditions present on the flame-treated sample surface. Furthermore, by observing the O(1s) signal, the quantitative increase of the surface oxygen by the flame-treatment can be determined.

As a result of the flame-treatment, the concentration of oxygen in the surface increases by a factor of 10 to about 8 atom% (fig. 5). This oxygen is largely due to C-O and C=O groups that have been formed in the surface.

4.1.1 Polyethylene (PE)

Fig. 6 shows the development over time of the total surface tension as well as of the polar and disperse component up to 3 weeks after a corona pretreatment. During the first few hours, the total surface tension drops only slightly from around 44 mN/m to around 40 mN/m and then remains constant. Similar behaviour shows the polar component, which drops from about 27 mN/m to about 24 mN/m. If the level of these surface tension values is compared with the values of the total surface tension of 21 mN/m (polar part: 2 mN/m) of untreated films, the surface tension – in particular the polar part – is drastically increased due to the corona treatment. During the observation period of three weeks, the strongest change occurs on the first day. Thereafter, the surface tension values remain practically constant, and tend to fall slightly after about two weeks.

4.1.2 Comparison with test ink measurements

Fig. 7 shows the comparison of the surface tension values (determined with test inks and with the contact angle method respectively) of differently pretreated films.

The contact angle method allows a clear differentiation of the films with regard to their surface tension values, in particular the polar component. The test inks, on the other hand, provide only slight differentiability, although in this case they were graded among each other in steps of 1 mN/m. The general problem of the use of test inks is based on the fact that they represent a mixture of two different liquids which have different polar and disperse components.

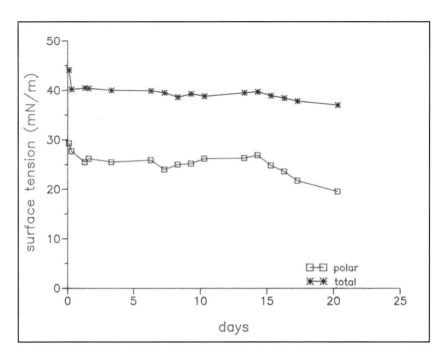

Fig. 6: Development over time of corona-pretreated PE films

Thus, the ratio of polar to disperse fraction for a series of test inks is never constant, but may change drastically from a test ink with smaller to a test ink with higher surface tension. Furthermore, evaporation and contamination effects can never be ruled out.

In addition to the theoretically inadequate assumption that spreading of the test ink on the solid surface occurs when the test ink and solid have the same surface tension values (neglecting of interfacial tension) and for the reasons given above, this method should be used only as the simplest method of operation.

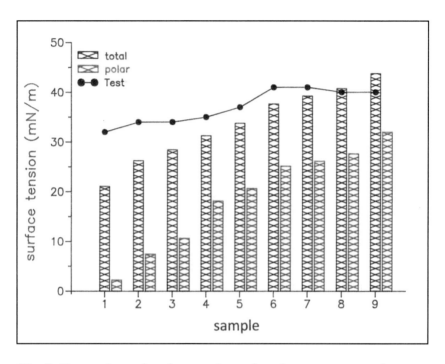

Fig. 7: Comparison of surface tension values from contact angle measurements with test ink results

4.1.3 Adhesion test on PE films

To check the quality of the corona pretreatment with regard to the adhesion of a water-based printing ink to the pretreated films, the so-called Tesa test was chosen. For the entire series of tests, only adhesive tape from one batch was used to exclude fluctuations in the adhesive force of the tapes (19 mm, transparent, Beiersdorf AG, Hamburg). The adhesive tape was applied to the printed film and pressed. The adhesive strip was then lifted at one end and pulled off quickly, jerkily. With this method, printed films that had been pretreated to different degrees were examined.

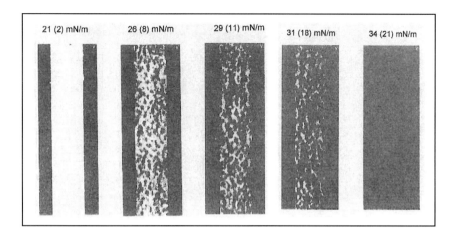

Fig. 8: Adhesion results of an water-based printing ink on corona-pretreated PE films

As can be seen from fig. 8, the weakest adhesion results in the film with the lowest surface tension and the best adhesion at high surface tension. The values over the test strips correspond to the values for the overall surface tension of the respective film (in brackets the polar component). The size of the exposed PE surface serves as a measure of the adhesion, i. e. no printing ink removal (no exposed PE surface) means best adhesion; correspondingly poor adhesion with large printing ink peeling off. In fig. 8, the exposed PE surface appears bright, the aqueous printing ink dark. Thus, in this case, a clear correlation between surface tension and adhesion can be observed.

4.2 Separation

One of the reasons for the occurrence of surface defects (craters) is the fact that, during film formation upon evaporation of the solvent, separation of the various resin components may take place if these components have different surface tensions [27, 28].

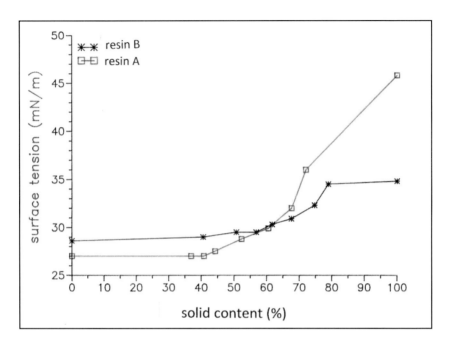

Fig. 9: Surface tensions of resins A and B as a funtion of solid content

To test this idea experimentally, two electrocoat resins (A and B) were investigated, which were known from the outset to be incompatible and cause craters. From these samples, the surface tension was now determined as a function of the solid content. The surface tension values of the resin solutions were measured with a tensiometer according to the so-called ring method. This measuring method fails from a solid content of the resin solution of about 70-80 %. The surface tension of the pure resin (100 % solid content) was now determined with the aid of the AURAM contact angle measuring device, after the resins had been knifecoated onto metal sheets (with subsequent oven drying). As can be seen from fig. 9, the difference in surface tensions between the two resins is very pronounced, and it has now been expected, according to the idea described at the outset, that the combination of these two resins will exhibit a surface tension substantially equal to that of the resin B

110

corresponds to ("floating" of the low surface tension ingredient). This idea was supported by the measurement, the 100 % value of the combination did not differ significantly from the value of the surface tension of the resin B.

4.3 Craters

A crater represents a material defect in the coating layer which either occurs during application or during film formation and can basically be described as a local wetting defect. For the occurrence of this wetting defect locally a surface tension difference between two substances or a surface tension gradient in the layer must be held responsible, which represent the driving forces for a material shift against the viscous forces in the liquid coating. If non-uniform evaporation of the solvent occurs in the wet coating film due to local temperature differences, the resulting surface tension gradient may be responsible for the formation of craters. A similar effect may be caused by local variations in the solvent concentration in the applied coating. The surface tension gradients in the wet coating film lead to local material movements in the coating film. A phenomenon also related to surface tension gradients during the evaporation phase with the formation of hexagonal flow patterns is known in the literature under the name Bénard's cells. To avoid craters, the surface tension at the air interface must be locally the same everywhere. Furthermore, a higher viscosity and a thinner coating layer counteract crater formation [29].

Overspray with low surface tension that gets on the wet coating film may be another cause for the crater formation. The spreading of these substances on the wet coating film is caused by surface tension differences between the coating film and overspray substance [30-32]. Foreign substances, e. g. dirt particles, appear as dirt inclusions, if their surface tension is greater than that of the coating, otherwise they produce craters. Contains already the liquid coating before application inhomogeneities with low surface tension craters can also arise. In addition to the listed problems, special attention must be paid to the area to be painted. No substances with low surface tension, e. g. grease, fingerprint, lubricant from the hanger, etc., have to be located on the surface. The problem with coating systems, which consist of components with different surface tensions, has already been explained in the previous section using the example of the incompatible electrocoat resins. A typical crater appearance is shown in figure 10 (light microscopy image [33]).

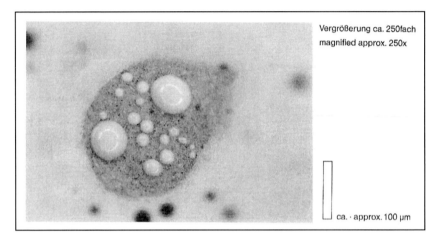

Vergrößerung ca. 250fach
magnified approx. 250x

ca. · approx. 100 μm

Fig. 10 : Surfacer crater (top view)

Countermeasures / reduction of crater fragility

Throughout the film-forming process, the surface tension of the wet paint film should be low and locally constant. A low surface tension can be reached by additives which themselves have a low surface tension and are incompatible with the paint. Because of these two properties, the additive is displaced from the paint and accumulates in the coating boundary layer. In this boundary layer, the additive must continue to be free to move so that it can diffuse through the new layer after applying a further coating layer and does not remain in the boundary layer between the two coating layers and leads to adhesion problems. Another problem can be caused by localized overconcentration of the additive, creating a localized area of low surface tension that will

112

not wet and thus promote cratering. The surface tension of the surface to be painted should be as high as possible. For this reason, careful pretreatment or cleaning of the surface should be carried out. By degreasing, grinding, blasting, phosphating surfaces are freed of substances and inhomogeneities, which usually have a low surface tension.

5 Summary

The wettability of a solid surface by a liquid is determined by the phases involved. While relatively simple methods are known for the liquid phase (e. g. ring method), the contact angle of several test liquids on the surface must be measured to determine the surface tension of the solid surface. The use of these methods for describing wetting and adhesion phenomena has been explained using examples from the fields of plastics coating and printing technology. The importance of surface tension in separation processes was demonstrated and the crater problem was discussed.

References

01.   T. Young, Phil. Trans. Roy. Soc., 95 (1805) 65

02.   W. A. Zisman, Adv. Chem. Serv., 43 (1956) 1

03.   D. K. 0wens and R. C. Wendt, J. Appl. Polym. Sci.,13 (1969) 1741

04.   D. H. Kaelble, J. Adhesion, 2 (1970) 66

05.   D. H. Kaelble, Physical Chemistry of Adhesion, John Wiley & Sons, New York (1982)

06.   P. L. du Noüy, J. Gen. Physiol., 1 (1919) 521.

07.   DIN 53914, DIN EN 14370

08. DIN 53593 (rejected)

09. A. W. Adamson, Physical Chemistry of Surfaces,
John Wiley & Sons, New York (1982)

10. DIN 53364

11. M. Osterhold, K. Armbruster and M. Breucker,
Farbe Lack, 96 (1990) 503

12. U. Zorll, Chem. Ing. Tech., 60 (1988) 162

13. M. Michel, Z. Werkstofftechn., 18 (1987) 33

14. N. Inagaki et al., J. Adhesion Sci. Technol., 4 (1990) 99

15. G. Gatenholm, C. Bonnerup and E. Wallström,
J. Adhesion. Sci. Technol., 4 (1990) 817

16. H. Gleich and H. Hansmann, Adhäsion (1991), No. 1-2, 15;
Adhäsion (1991), No. 3, 27

17. R. Foerch, N. S. McIntyre and D.H. Hunter,
Kunststoffe, 81 (1991) 260

18. G. Menges et al., Kunststoffe, 80 (1990) 1245

19. K. Armbruster and M. Osterhold,
Kunststoffe, 80 (1990) 1241

20. M. Osterhold and K. Armbruster
Farbe Lack, 97 (1991),780

21. K. W. Gerstenberg, Coating, (1990) 260

22. M. Osterhold and A. Wiechers,
Materialprüfung, 35 (1993) 178

23.   A. R. Blythe el al., Polymer, 19 (1978) 1273

24.   D. Briggs and C. R. Kendall, Int. J. Adhesion and Adhesives, (1982) 13
.
25.   D. Briggs et al., Polymer, 24 (1983) 47

26.   M. Strobel et al., J. Adhesion Sci. Technol., 6 (1992) 429

27.   Synres NL bv, Pigment and Resin Technol., (1973) 13

28.   U. Zorll, Farbe Lack, 95 (1989) 801

29.   C. M Hansen and P. E. Pierce, lnd. Eng. Chem. -Pred. Res. Dev., 13 (1974) 218

30.   G. P. Bierwagen, Prog. Org. Coat., 3 (1975) 101

31.   F. J. Hahn, J. of Paint Technol., 43 (1971), No. 562, 58

32.   B. Jacobson, Polym. Paint Col. J., (1978) 701

33.   P. v. d. Kerkhoff, Fehlstellenkatalog, Herberts, Wuppertal, Germany (1990)

34.   M. Osterhold, Farbe Lack, 99 (1993) 505

Chapter 10 – Coating Defects

Microscopical Defect Analysis

1 Introduction

Substrate defects (underground defects), inhomogeneities in the liquid coating or due to improper application can cause a wide range of defects, such as craters (small unwetted areas), boils (bubbles, needle-like holes) or dirt inclusions may occur in the coating surface. These defects can be documented in many cases by light microscopy, but a clear determination of the cause of the defect is in some cases difficult and requires a lot of skill and experience in sample preparation.

2 Investigation methods

The first analytical assessment of a defect is usually done in light microscopic top view, for which no complex sample preparation is necessary. Often, however, the cause of the defect can not be determined. For further investigations essentially the following two methods of sample preparation are used.

During preparation in the so-called cross section, the sample is fixed edgewise in an embedding material (e. g. casting resin), and then sanded with various abrasive papers to the centre of the defect to be investigated. For a thin section (microtome cut), the sample is fixed in a clamping block, and a thin slice of the sample is planed off with the aid of a knife. When using fully automatic microtomes, the layer thicknesses are about 2 to 10 µm. For the exact assessment of coating layer thicknesses, the cross section technique is generally better suited, as it can be observed really perpendicular to the sample surface. Cross sections can only be viewed in reflected light. For transmitted light investigations, microtome sections are the method of choice, which, depending on the problem, can be performed parallel or perpendicular to the substrate surface.

The resolution limit of light microscopic methods is about 1 µm. It depends on the wavelength of the light used and can be partly improved even further by use of UV light in modern microscopes.

In addition to the light microscopic investigations, topographic analyses with a resolution in the nanometre range or element analyses in the SEM (Scanning Electron Microscope with energy- or wavelength-dispersive X-ray analysis EDX/WDX) can also be carried out. For the determination of the cause of a defect, however, a mass spectroscopic examination is often necessary. Apart from LAMMA (Laser Microprobe Mass Analyzer) the analysis with ToF-SIMS (see below) plays an important role.

In laser microprobe mass analysis, a short laser pulse is used for the formation of ions at the target surface. The enormous power density

118

results in the evaporation and partial ionization of target material in a diameter of 3 to 5 µm and a depth of some tenth of a micrometer. The generated ions are analyzed in a time-of-flight mass spectrometer. Due to the high pulse energy, however, nearly all organic molecules are completely destroyed. Therefore, the main information obtained by LAMMA is the element composition of the analysed target area, with significantly higher detection sensitivity than EDX.

In time of flight secondary ion mass spectroscopy (ToF-SIMS), a pulsed primary ion beam is used for the formation of secondary ions at the target surface, which are analysed by a time-of-flight mass spectrometer. The main advantage of ToF-SIMS is the semi-quantitative characterization and identification of the organic and inorganic composition of the topmost monolayer at the target surface, with high sensitivity.

ToF-SIMS, on the other hand offers the ability to distinguish between different silicone oils, which are among the main causes of defects. In the spectra of negative ions, silicone oils have a characteristic fragmentation pattern. In the mass range below 100 u (m/z) one or more fragments can be observed. In the higher mass range silicone peaks of the fragments with additional repeat units of the polymer occur. These reiterated signals can be used for the identification of the polymer repeat unit of the specific silicone oil. With this tool it is possible to distinguish silicone oils, based on different monomers. It can also be observed, that PDMS based silicone oils are able to form up to three different fragmentation series in the spectra of negative ions. A significant change in the ratios between these silicone oil fragmentation series inside of a defect compared to the surrounding intact surface indicates a change in the silicone oil composition.

Possible reasons for such a change are the contamination with a foreign silicone oil, or, for coatings with more than one silicone additive, the separation or enrichment of one of them.

3 Practical examples

3.1 Solvent boils

An example of the difficulty in identifying the cause of a defect in the light-microscopic top view is the examination of boils or pinholes. Fig. 1 shows a typical pinhole in top view. In the centre of the boil, the clearcoat does not seem to be closed, but it is unclear whether there is a disturbance even in the underlying basecoat. Only in the cross section (also fig. 1), the cause of this boiling defect becomes clearly. Caused by an excessively high layer thickness of 96 µm (instead of 10 to 20 µm), a cooking blister has been formed in the basecoat due to solvent evaporation. This then continues in the clearcoat and thus leads to a defect on the surface.

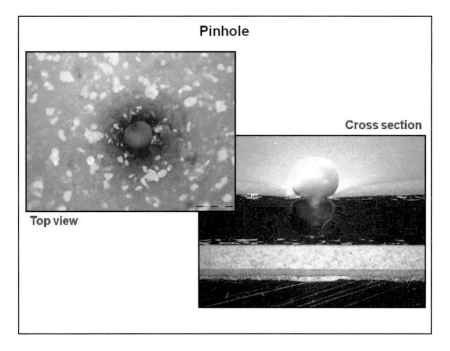

Fig. 1: Pinhole/solvent boil

3.2 Craters

One of the most difficult problems is the analysis and root cause of craters. They represent areas that are poorly or not wetted by the liquid paint, with typical diameters ranging from fractions to a few millimetres.

Apart from impurities in the liquid paint, the cause of such disturbances lies in local variations of the wettability of the substrate, caused, for example, by contamination or accumulation with substances (e. g. silicone oils, grease) which themselves have a too low surface tension and thus can not be wetted by the liquid paint.

Overspray with low surface tension that gets on the wet coating film may be another cause for the crater formation. The spreading of these substances on the wet coating film is caused by surface tension differences between the coating film and overspray substance. Foreign substances, e. g. dirt particles, appear as dirt inclusions, if their surface tension is greater than that of the coating, otherwise they produce craters. Contains already the liquid coating before application inhomogeneities with low surface tension, as impurities, craters can also arise.

In addition to the listed problems, special attention must be paid to the area to be painted. No substances with low surface tension, e. g. grease, fingerprint, lubricant from the hanger, etc., have to be located on the surface. By degreasing, grinding, blasting, phosphating surfaces are freed of substances and inhomogeneities, which usually have a low surface tension.

Fig. 2 shows the microscope image of a production line crater in a complete coating system. In the centre of the defect, which is reaching down to the basecoat, a hole can be observed even in that layer. This crater was analysed by ToF-SIMS in comparison to the surrounding intact clearcoat surface.

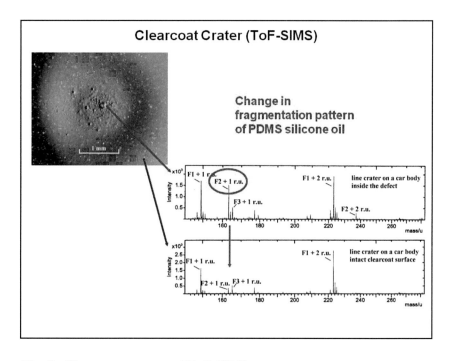

Fig. 2: Clearcoat crater and ToF-SIMS spectra

In the spectra of negative ions (also presented in fig. 2) a significant change in the fragmentation pattern of PDMS based silicone oils occurred between the crater and the intact clearcoat surface. This led to the assumption, that the crater was caused by a silicone oil on PDMS base.

In fig. 3, the microscope image (top view) of a primer crater on a car body is shown. First LAMMA analysis was carried out on two similar craters from the same panel. By this method it was not possible to find any indication to a possible crater cause.

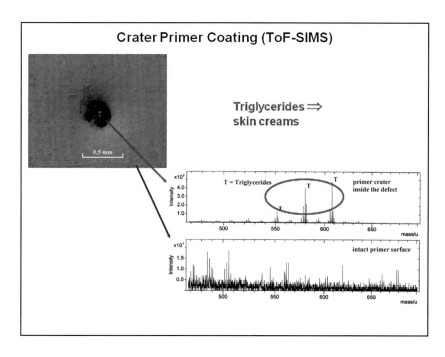

Fig. 3: Crater primer coating and ToF-SIMS spectra

Then the crater of fig. 3 was analysed with ToF-SIMS and significant differences between the defect and the intact primer surface were found. In the spectra of positive ions (spectra also see fig. 3) additional signals of triglycerides and in the spectra of negative ions, the corresponding fatty acid fragments occurred. Triglycerides are often used in skin creams. Therefore, the crater for example might be caused by a fingerprint of an employee in the paint shop of the car builder.

3.3 Overspray

In addition to causing craters, overspray can also lead to pin-like defects, gel-like foreign particles, may be confused conceivably with inclusions of real particles.

123

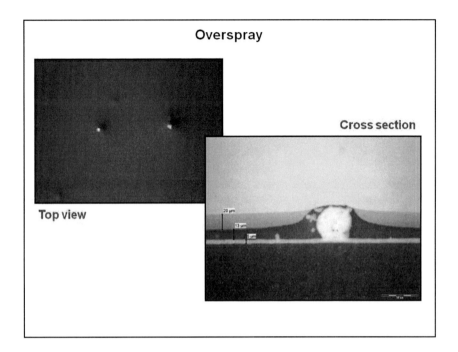

Fig. 4: Overspray

An example of this is shown in fig. 4 in the microscopic top view (left image 1 cm approx. 100 μm). In the cross section (also fig. 4) it can be seen that the defects are overspray of the primer, which can not be covered by the overlying layers.

3.4 Fibres

Even the smallest fibres can also lead to coating defects. To prevent and combat these defects, it is important to determine the type of fibre and thus its possible origin. Fig. 5 shows a fibre enclosed in a coating layer. In the cross section (also fig. 5), it can be seen from the typical bean-shaped form that this is a cotton fibre, which could descended e. g. from lab coat material.

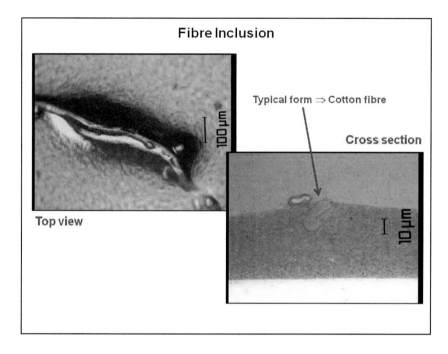

Fig. 5: Fibre inclusion

3.5 Particles

Another class of defects that are often studied are particles in surfaces. For inorganic characterization, analysis with LAMMA is particularly suitable for this purpose. A simple but not trivial example is shown in fig. 6. On the microscope image a small ball of metal can be observed on the top of an E-coat layer.

In the LAMMA analysis of the particle iron and copper were detected in the positive ion spectrum. In comparison to that, the E-coat only showed signals of the pigments (Al-silicate and Ti-oxide) and the organic resin matrix. With this optical and mass spectroscopic information the conclusion was drawn, that the defect was a welding pearl, because the wire used for electric welding consists of iron, which is covered by a copper layer to avoid corrosion.

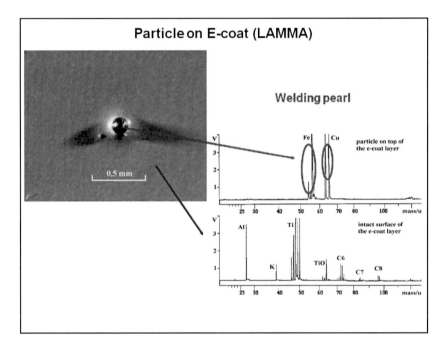

Fig. 6: Particle on E-coat

The examples presented here from the field of automotive coatings represent only a small selection of surface defects. A large number of defects is described on the basis of light microscopic images for example in [1, 2].

References

01.   Fehlstellen-Katalog – Catalog of Defects (2. Aufl.), überarb.
        von T. Haase, M. Osterhold, Herberts, Wuppertal (1996)

02.   P. v. d. Kerkhoff, H. Haagen, Lackschadenkatalog,
        Vogel-Verlag, Würzburg (1995)

Chapter 11 – Weathering

Introduction to Weathering Testing

1 Introduction

Coatings, as other organic materials (e. g. plastics), are damaged by sunlight, heat, rain or moisture and change their performance characteristics over time. The stress caused by these climatic factors is referred to as "weathering", in this context the change in coating properties over time is often described as "aging". It is to distinguish between two fundamentally different processes: the initiation of aging processes by solar radiation as the primary reaction and the additional influence of aging processes due to further secondary reactions by the other climatic factors (temperature, humidity, etc.).

2 Influencing factors

The most important weathering-related degradation phenomena of coating systems are largely determined by the following factors:

- Solar radiation
- Temperature (increased, low, cycles)
- Water
  - in solid form (snow, ice)
  - liquid (e. g. rain, condensation)
  - gaseous (e. g. high relative humidity)
- Normal air components such as oxygen, ozone, carbon dioxide
- Air pollutions as
  - gases (e. g. nitrogen oxides and sulfur oxides)
  - mist (e. g. aerosols, salts dissolved in water, acids and alkalis)
  - solids (e. g. sand, dust, dirt)

The effective proportion of the spectrum of extraterrestrial electromagnetic radiation (solar spectrum) on earth is the so-called global radiation in the wavelength range from about 300 nm to 2500 nm.

Their intensity is influenced by the latitude or the position of the sun (equator, north pole), the elevation (increasing UV content with increasing elevation), local climatic conditions (cloud cover, humidity, air pollution) and by season and daytime.

Since the binding energy of common chemical bonds is in the range of the quantum energy of the UV radiation, the UV range is mainly responsible for the initiation of the photochemical primary processes of the aging of polymeric materials. In addition, especially in filled coating systems, the visible spectral range is important because pigments, fillers or certain additives can absorb in the visible region.

3 Outdoor weathering

In comparison to Central Europe, the degradation processes in the tropics or subtropics (e. g. Florida) are about two to three times faster. The reasons for this are in particular the differences in the global irradiation, the air temperature and in the humidity as influencing weathering factors. Thus, the colour-dependent temperature, which sets in the coating film, as well as moisture and oxygen, determines the speed of the chemical degradation processes taking place.

Typical values of the temperature can be, for example, on the body shell of cars up to 70 °C, in the interior tray above 90 °C.

Water can affect the coatings as rain, dew or moisture. Nocturnal condensation phases are typical for the subtropical climate. The purely mechanical stress caused by swelling and de-swelling can lead to tensions, which ultimately lead to cracking in the coating. Also, by chemical interaction of the coating with the water (hydrolysis, radical formation) damage can occur.

For the testing of coating systems for practical use two climatic zones have been established as reference climates; Florida with humid-warm and Arizona with dry-warm climate. The long-term weathering test in Florida (fig. 1) is mandatory, for example, for all automotive topcoats.

Options to carry out an accelerated load for automotive coatings can be obtained with sample exposition at so-called black-box weathering in Florida. For this special, very demanding weathering a heat accumulation is produced under the panels, similar to that in the passenger compartment or trunk of cars.

Typical weathering times are between 12 and 36 months, sometimes up to 120 months.

Fig. 1: Florida weathering, exposition type open rack (45° south)

During and after completion of the weathering important coating properties such as gloss and colour are measured; blisters, cracks, water spots or chalking are assessed visually. To check the adhesion properties the cross-hatch is performed.

In another outdoor exposure method (EMMA), which is used specifically in Arizona, the sunlight is bundled over 10 mirrors directed to the samples. The test bench is automatically aligned with the sun again and again so that maximum solar radiation is guaranteed. As a result, about eight times the intensity of sunlight compared to Florida can be achieved. For the purpose of regulating the sample temperature, the samples are cooled from the back with air. In order to achieve a better correlation to the moist Florida climate, the samples can also be sprayed with de-ionized water ("Emmaqua" weathering).

4 Accelerated weathering

In the case of accelerated weathering in devices (artificial weathering), the following points are important

- Colour change, cracking during weathering
- Relation of device and outdoor weathering (time-reduction, predictability) and
- Product comparison

In the case of xenon radiators whose radiation (emission spectrum) comes closest to the global radiation in the UV range important for changes in the coating binders, the use of corresponding filters yields a very good approximation to natural solar radiation (see fig. 2). Other devices with fluorescent lamps produce a UV-rich light for testing especially clearcoats.

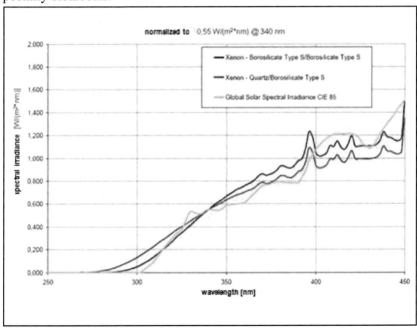

Fig. 2: Sunlight and xenon light with different filters

Fig. 3: Arrangement of the test panels in the sample chamber

Radiation sources as well as filters age, i. e. the energy that they put to the sample undergoes changes over the course of their practically proven lifetime, which vary depending on the wavelength range. This requires an automatic controllability of the light intensity impinging the samples.

In addition to the light exposure, in the devices also the humidity, irrigation and temperature program are controlled. The relative humidity is during the dry time in the devices at approx. 50 %, regulated to about 95 % in the time of irrigation. Irrigation is done with de-ionized water to prevent staining and sedimentation on the samples from the beginning. It is usual to change between irrigation and drying phases (see, for example, [1, 2]).

The heating of the samples in the interior of the devices is monitored with the so-called black panel thermometer, whose measured value must not exceed a predetermined temperature. Fig. 3 shows the sample chamber of a modern accelerated weathering device.

Fig. 4 shows the crack formation of a clearcoat system and the colour change of a red topcoat system after weathering [3].

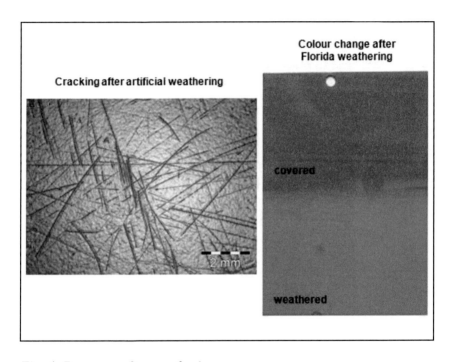

Fig. 4: Damages after weathering

With these measures, uniform conditions are created during weathering and reproducible results are obtained with weathering times of several hundred or thousand hours. Depending on the accelerated weathering method and customer requirements, typical weathering times are between 2000 h and 5000 h.

By varying the radiation intensity and duration, temperature or humidity cycle, it is basically possible to change the time reduction in the weathering devices. DIN EN ISO 4628 deals with the assessment of coating damage that can also be applied to aging after weathering. A summary on weathering procedures can be found e. g. in [4]. There are existing also various DIN and company standards for accelerated/artificial and outdoor weathering.

References

01.   SAE J 1960 (rejected)

02.   SAE J 2527

03.   U. Schulz, Kurzzeitbewitterung, Vincentz, Hannover (2007)

04.   M. Osterhold and D. Kegelbein, in: Band III HighChem hautnah, GDCh (2008)

Literature sources

The individual chapters are based in total or partly on the following publications in journals or on conference/seminar papers:

Chapter 1: Introduction Rheology

M. Osterhold
„Viskosität auf dem Prüfstand"
Farbe Lack, 123 (2017) 42

M. Osterhold
„Rheologische Charakterisierung von Lacken"
Vortragstagung der GDCh-Fachgruppe
Lackchemie, Paderborn (2016)

Chapter 2: Rheological Measurements

M. Osterhold
„Improving Rheology Measurement"
European Coatings J., (2016), No. 11, 44

Chapter 3: Application Examples

M. Osterhold
„Rheological Methods for Characterising Modern Paint Systems"
Prog. Org. Coat., 40 (2000) 131

Chapter 4: Rheology and Surface Charge

M. Osterhold
„Charakterisierung disperser Systeme"
Farbe Lack, 101 (1995) 683

Chapter 5: Particle Size Determination

M. Osterhold
„Charakterisierung disperser Systeme"
Farbe Lack, 101, 683 (1995) 683

Chapter 6: Thermal Analysis

W. Schlesing, M. Buhk, M. Osterhold
„Dynamic mechanical analysis in coatings industry"
Prog. Org. Coat., 49 (2004) 197

M. Osterhold
Manuskript DFO-Seminar „Industrielle Lackiertechnik", (2002)

Chapter 7: Scratch Resistance

M. Osterhold
„Marking Time"
European Coatings J., (2018), No. 01, 52

Chapter 8: Surface Structure

M. Osterhold
„Patterns of Roughness"
European Coatings J., (2016), No. 07/08, 44

M. Osterhold, K. Armbruster
„Oberflächenstrukturanalyse an Substraten und Lackierungen -
Metall, Kunststoff, Lack"
DFO-Technologie-Tage, Berlin, Tagungsband, (2005) 4

Chapter 9: Surface Tension

M. Osterhold
„Bedeutung der Oberflächenspannung für physikalische
Lackeigenschaften"
Farbe Lack, 99 (1993) 505

Chapter 10: Coating Defects

M. Osterhold, E. Frigge
„Mikroskopische Fehlstellenanalyse"
Band III HighChem hautnah, GDCh (2008)

Chapter 11: Weathering

M. Osterhold, D. Kegelbein
„Bewitterung"
Band III HighChem hautnah, GDCh (2008)

Biography

Dr. Michael Osterhold studied physics and received his Ph.D. at the Ruhr-University, Bochum (Germany). After several years as researcher and lecturer, he joined Herberts (Wuppertal) in the late 1980s, later DuPont Performance Coatings. He was responsible for the unit R&D-Services EMEA, incl. e.g. the Physics, Materials Testing, Weathering and Analytical Science departments. In 1995 he was recipient of the Farbe&Lack-Prize. He is author of about 70 scientific/technical papers. Since 2011 he has been working as a scientific consultant and as an adjunct Professor of Physics at an University of Applied Science.